Praise for
The Clockwork Girl

'Historical fiction with a fantastical twist, done with
verve and skill'
IAN RANKIN

'An atmospheric and constantly surprising thriller'
THE TIMES

'Evocative, chilling, compelling. Tremendous'
TAMMY COHEN

'Spellbinding, gripping, immersive and deliciously gothic.
No one gets under history's skin like Anna Mazzola'
ERIN KELLY

'Dark and bewitching. I loved it'
LAURA SHEPHERD-ROBINSON

'A deliciously dark historical novel of thrilling originality.
Immensely clever and entertaining'
ESSIE FOX

'It's rarely I come across a book that astounds me. [This] is
one such book. It is breathtakingly good'
ABIR MUKHERJEE

'Unbelievably immersive. A gripping, gothic tale . . . three
unforgettable women caught in a web of intrigue and
murders. You really won't want to miss this'

'A dark delight. A gothic gem'
BETH UNDERDOWN

'This intricate and twisting story kept me guessing until
the end. An absolute masterpiece!'
JENNIFER SAINT

'A thrilling tale of intrigue, espionage and survival.
Exquisitely written, brilliantly atmospheric . . . a
masterpiece of suspense and sensitivity'
MARY CHAMBERLAIN

'Deliciously creepy, alive with the rustling silks, rotten
smells and lethal egos of a divided Paris. Loved it'
SARAH HILARY

'A dark tale of vanished children, an obsessive genius,
and courtly machinations. Another accomplished slice of
historical noir from a sure-footed exponent of the genre'
VASEEM KHAN

'A superb historical novel set in a brilliantly evoked 18th
century Paris, *The Clockwork Girl* will delight and enthral
and leave you wanting more'
WILLIAM RYAN

Vivid and meticulously detailed. Anna Mazzola has
crafted this dark jewel of a novel with the precision of a
watchmaker . . . Glittering macabre'
KATE GRIFFIN

For my sister, Laura.

One

Paris

I

Paris, 1750

Madeleine

Today was the day Maman priced up the girls. Best on such days to slip away. That was why Madeleine now walked past the slaughterhouse, where the blood had congealed into a dark gash across the snow and where carcasses hung from hooks, pale arses to the morning sky. In the glassy air her ungloved hands smarted, the skin of her knuckles cracked and raw. Not much of a day to be out for a walk, but damned if she was staying home to listen. Besides, there was something she needed; something that couldn't wait.

Madeleine turned off the Rue Pavée to enter the labyrinth of the Quartier Montorgueil, the alleys too narrow, the houses too high, so that the sun was kept out and the stench kept in, the streets dark and rank as the devil's *connard*. Ancient buildings leaned into one another like crowded teeth, their crumbling brickwork patched together, their windows stuffed with rags. Now and again a face emerged from the shadows: a child, like as not, with the tell-tale features of hunger, generations deep. Better here, though, among the lowest of the low – *le bas peuple*, as they called them; the slum-dwellers and doorway-lurkers, the homeless and the shoeless – than at

Maman's so-called 'Academie', where the monthly inventory would be in full swing.

It was a crying shame, her mother always said, and truly she took no pleasure in it, but she was running a business and human flesh was a damned changeable thing: breasts ripened or withered, diseases took root, skin stretched or pitted, sores filled and burst. Babes – despite a barrage of precautions – were wont to begin and blessed difficult to get out. And once in a while something would happen, just as it had happened to Madeleine, to halve a girl's value in the space of one day. There'd always be at least one girl – knocked up or knocked down – who would, in Maman's phrasing, be put out to pasture. Only there was no pasture in the backstreets of Paris. There was a black river of refuse, broken bottles, fish heads. Right now in January there was sleet and snow, blunt figures huddled together in doorways, and the occasional stiffened corpse.

Reaching the Pointe Saint Eustache, Madeleine emerged into the powdery winter sunshine, the air dulled by smoke and ash. She'd left the environs of the poorest of the poor and reached the purlieus of the merely wretched. She skirted the rowdy market of Les Halles and walked south towards the Pont Neuf, sedan chairs keeping the bejewelled and furred bobbing above the thinly-clad poor. As she crossed the Rue Saint-Honoré, a gilded carriage flashed past her, striking sparks from the paving stones, a glimpse of a satin-swathed woman behind the glass. Might be an aristo, might be a *femme entretenue* in the carriage of her keeper – the only way for a girl to grow rich in Paris was lying on her back.

Finally she was at the river, from where she could see the spires of the Cathédrale Notre-Dame reaching into the winter skies, the single iron spike of the Sainte-Chapelle. Looking only at the skyline, you could imagine that Paris was a rich

4

place, a beautiful place, a city of learning and piety. No doubt it was in part, but not the parts Madeleine knew, for they were the bits Paris kept buttoned. She walked on, along the banks of the dirt-grey Seine, until she reached the *apothicaire*'s shop, where bottles gleamed in the window like jewels. She hesitated a moment, steeling herself, then pushed at the oaken door.

Inside, the air was spice-scented and warm, though the reception she received was cool. Two women were at the counter, talking intently, glancing at Madeleine and dismissing her as tat without even drawing breath. The apothecary himself was weighing a blue-coloured powder on his great brass scales and took no notice of her at all. As she waited, toes thawing painfully in her boots, Madeleine stared up at the rows and rows of glass jars and porcelain chevrettes lining the shelves. Some of the names were familiar to her: cloves, borage, comfrey, angelica. Some meant little: jalap root, cinchona, sarsaparilla; a box labelled Campeche amber.

'And she never arrived, would you believe it?' one of the women was saying to the other. She was, Madeleine thought, one of those sharp-edged women who'll find the worst in everyone.

'What happened to her?'

'Well, no one knows for sure. But there was that travelling fair that upped and moved on the very next day. Makes you wonder, doesn't it? I wouldn't be letting my daughter out on her own at that age, of course.'

The apothecary looked up, his eyes resting on Madeleine for a moment like black flies, then darting away. He wrapped the women's purchases, took payment and stretched his thin lips into a smile. As soon as Madeleine walked forward to the counter, the smile vanished.

'No improvement?'

'Very little.'

'You did as I advised?'

'I did.' She paused. 'Is there something else that might help?'

The man considered her. 'Possibly, but it'll cost you.'

Didn't it always? For an instant she considered pleading with him, telling him of their circumstances, but she knew there was little point. No one gave anything in Paris for free, certainly not to a girl like her.

'More than last time?' she asked.

'I'd say so, yes. The stuff is expensive, brought over from the Americas.'

Madeleine raised her eyebrows. 'Exotic. I see.'

She watched in silence as he pounded the ingredients in a pestle and mortar, crushing the seeds to dust. A powerful aroma filled the air, nutmeg mixed with something burning and bitter. The apothecary transferred the medicine to a green glass vial and set it down on the counter before her. 'Two louis, it's worth.'

Madeleine's eyes were on his veined white hands. 'Very well.' She looked back at the door. No other customer had entered.

The apothecary walked over and turned the sign, indicating the shop was shut.

Madeleine walked back home slowly, her cloak wrapped tightly about her, passing women in great fur mufflers walking tiny dogs on leashes, and servants out on errands, raw-faced in the cold. She made her way up the Rue de la Monnoye, looking into the windows of milliners' shops filled with feathered hats, like a flock of exotic birds. Is this how her father's parrots had ended up, she wondered – made into bonnets for the *gens de qualité*? He'd been a *oiseleur*, her papa, selling birds and other small animals, from a shop on the Quai de la Mégisserie. After his death, Maman had sold his stock to another trader

6

– cockatoos, finches, white mice and squirrels, all boxed up and passed on quicker than you could say jackanapes. The rent on the shop being long overdue, they'd cleared out quick from the house she'd grown up in and set up in the Rue Thévenot, selling a different kind of bird.

She walked on, past jewellers' shops glimmering with sapphires and rubies, or paste that looked very much like them. That was the thing with Paris, you had to learn the trick of telling what was real and what was false. Gems, hair, cleavages, characters – all could be easily faked. *Smallpox taken your eyebrows, Madame? Buy this pair of finest mouse fur. Lost your teeth in a brawl, Monsieur? Have a set drawn from another's grave.* Madeleine could see her breath, a puff of white in the icy air, and her hands were now almost numb. Still she didn't hurry, for she doubted her mother would be done. Genevieve Chastel – 'Maman' to the girls – was nothing if not rigorous.

Some of the ones her mother put out would survive, at least for a time, on the backstreets as *filles publiques*, the lowest in the pecking order of the many who sold sex in the demi-monde, running from the *femmes de terrain* in the public gardens to the bejewelled mistresses of Versailles. Some of Maman's girls would die nearby in the filth of the Hôtel Dieu, the hospital God had abandoned long ago. Some would be swept by the police into the countryside. There was little point in thinking of it. If she did, she'd only be reminded of how chancy her own state was, how close she was to the brink. She'd no money of her own, no property, no references. If Maman put her out, she'd be on the streets like the rest of them, more mutton for the Paris pot. That was why it was best never to get too close to the other girls. She tended to them as required – brushing and plaiting their hair, pinning them into their dresses, changing their spunk-stained sheets. But there was no value in trying to help them. She had to focus on saving herself and, more importantly, Émile.

'Aren't you a cold one?' her elder sister, Coraline, had said earlier as she headed for the door. Rich, of course, coming from her.

Well, better to be hardened to a chip of jade than crushed, like the others, to dust.

It was past ten o'clock by the time Madeleine turned into the Rue Thévenot, her home since the tender age of twelve. (Tender, in fact, as a prime cut of beef, and sold for not much more.) The narrow street had been swallowed in shadow, all the better to hide the flaking house-fronts and rubbish-strewn ground. As she passed the vinegar-maker's, the tang of fermenting wine in the air, her eyes sought out the hunched figure of the girl who'd been living in a doorway these past few weeks. Sure enough, there she was, a man's coat tied about her with string, a coat that'd been perhaps her father's or brother's. Madeleine hadn't asked, had barely spoken to her. The girl's story would be much the same as all the other orphans and outcasts living in the city's stinking streets – disease, debts, liquor, death – and Madeleine had no wish to hear it.

'Demoiselle?' the girl stretched out her grubby hand.

Giving a tight nod, Madeleine reached into her pocket and took out the last few sous she had. Not enough for much, but maybe enough for some soup.

'Thank you. You're very kind.'

Very stupid, more like. Madeleine thought to herself as she continued up the road. The girl wouldn't survive a winter like this, so why prolong her suffering? She stopped. An icy breath seemed to touch the back of her neck. But when Madeleine turned she saw only a thin black cat, its eyes glinting in the shadows.

The *Academie*, as Maman called it, comprised the middle two floors of a tall, soot-stained building at the end of the street, bowed and blackened like a filthy finger, beckoning customers to the door. There was a staircase at the back by which the punters could enter and, though there was no sign marking the brothel, the place could be recognised at ten paces by its customers' distinctive stink. The men marked their territory like tomcats and the steps reeked with accumulated piss. Breathing only through her mouth, Madeleine entered the house, shut the door quietly, took off her muddy boots, and crept up the carpeted stairs. Behind a closed door she could hear a girl weeping and she walked quickly and noiselessly past.

'Where've you been?' A small figure jumped up as she reached the landing: her nephew Émile, face unwashed, hair unbrushed. 'You've been hours and I'm hungry.'

Madeleine stared at his grimy face with a mixture of love and annoyance. 'Did no one think to give you breakfast? No, of course they didn't. Why feed an eight-year-old boy? Go clean your face and hands, you little horror, and I'll get us both something to eat.'

The house was quiet, she noticed, as she warmed milk in a pan and sliced a loaf of bread; a strained sort of silence, of held-in anger and whispers behind bolted doors.

They made faces at each other as they sat at their breakfast, Émile snorting milk back into his bowl at Madeleine's impression of Grandmaman. After a minute, however, he grew serious. 'There's two girls gone this time. Odile and Lisette.'

'That so?' No surprises there. Odile hadn't had her courses these past two months and Lisette had for days been wearing lace gloves to hide the familiar rash of the pox. Madeleine stared at her own hands, the nails broken, the knuckles red, and tried not to think of their faces.

Émile's body was for several seconds wracked with coughing. She rubbed his back; thought of the medicine bottle, the man's cold, insistent fingers.

'Will she get rid of me, Madou?' he said, wiping his mouth. 'Will she make me go one day too?' It was a question he asked Madeleine often, and her answers were always the same.

'No, of course not, *mon petit*. You're her grandson.' Not that that meant much to her mother who, above all things, was a cold and clear-eyed businesswoman. 'And you're a useful little machine, aren't you? Always running errands and helping out.'

Mostly Émile was tasked with trailing punters back to their homes. There were plenty of slippery coves who gave a false name and false address, and it was in Maman's interests to follow them to their lairs. If a cull didn't pay, or if a cull caused problems, she needed to know where to find him. It was dangerous work for a boy like Émile, though. Dangerous, in truth, for anyone.

'And she won't turn you out, will she, Madou? She won't ever make you leave?'

Madeleine tensed and smiled to conceal it. 'Like I always say, Émile: I don't think so. I have my uses too.' Skivvy, teacher, voyeur, whore. But with Maman, there were no guarantees.

Footsteps in the hallway and then the door opened a crack, Coraline's painted face appearing from behind it. 'There you are, Madou. Maman wants you in the parlour. Now.'

Madeleine's heart plunged. 'Why?'

'Just come, will you? Quick.'

Madeleine felt Émile's eyes on her face, felt his fear in her own chest, and winked at him. 'Don't worry, *mon petit*. You know what she's like – probably wants me to massage her gnarly old feet. You finish eating your breakfast.'

But as she walked down the corridor, smoothing down her skirts, fear corkscrewed up through her chest; perhaps the inventory wasn't finished at all and she was the next to be priced.

As soon as she entered the faded parlour, she saw that her mother was dressed for company, the ceruse laid thick on her flaccid cheeks, her smile a slash of vermillion on white. Madeleine felt a queasy uneasiness as she saw that seated across from Maman, on the worn ottoman, was Camille Dacier, the man who'd been Suzette's least favourite client – a remarkable achievement in a rich and diverse field. It wasn't that he was ugly either. He had a sharp, roguish sort of face and unnerving eyes of different colours: one brown, one pale blue. Eyes that he now turned on her.

'Ah, the report writer.' He exuded an air of assumed authority, of contempt. When she'd first met him, Madeleine had known him at once for what he was: the worst kind of policeman.

'Go get the chocolate now, Coraline,' her mother said, patting her sister's hand. 'It'll be good to drink chocolate together.' In her taupe gown Maman looked like an oyster, fleshy and sickly pale. A pause and then Coraline moved towards the door, her skirts rustling in the silence.

'I must say you're looking well, Monsieur,' Maman said, turning to Camille. 'You're in good health?'

He didn't look healthy to Madeleine. His skin had the greyish tinge of a man who slept too little and drank too much. She recognised it from her father.

'I am, Madame. I've no complaints in that regard. Your girls are good and clean.'

Madeleine's throat felt dry, her palms slippery. Why had they brought her here?

'Oh, all my fillies are good girls, Monsieur, in their different ways. Madeleine here has always been the cleverest. The most noticing sort. Always knows what's o'clock, eh, Madou? Almost makes up for the scar.'

Maman didn't think that, of course. '*Une fichaise,*' her mother often called her: a thing not worth a curse.

Maman talked on for some minutes about the other girls she'd taken under her wing, to save them all from penury. Madeleine kept her expression as blank as a button. Listening to Maman talk, you'd think she was running a sanctuary for provincial girls, not a buttocking shop for whoremongers. 'The Academie', indeed. The only things she'd learnt here were the arts of dissembling, deceit, dulling your feelings and giving the perfect *pipe*.

Maman returned her gaze to Madeleine. 'Monsieur Dacier has an offer to make you, Madeleine. It's a good offer.' The smile was still fixed on her face as if applied there like her rouge.

'The reports you write for us,' Camille said, 'the tales you tell. They're very ... helpful.'

Madeleine gave a nod of acknowledgement. Her father had taught all three of them to read and write, after a fashion, and this was the use it was put to. For over a year now she'd been sending reports to Inspector Meunier, patron saint of brothels, protector of morals: the man who happily turned a blind eye to the selling of twelve-year-olds, the spread of venereal disease by *dames entretenues*, the frauds committed by *macquerelles*, but who was oh so very eager to know about the predilections, perversions and wheedled-out confidences of the culls who came to the Rue Thévenot. Most aristos went to the more genteel *sérails* in the Rue Denis or kept a mistress or two, but there were some who liked to slum it, and when they did, Madeleine noted it all down. Politicians who liked nothing better than for a girl to piss in their face; clergymen partial to a 'virgin' child. They should know better, for the state was always watching – Paris was criss-crossed with a network of *mouches* – and every institution had its spies.

'Well,' Camille said, 'now it's time for me to tell you a story, Miss Chastel.' He leant forward and lowered his voice, fixing her with his unnerving eyes. 'In a tall house on the Place Dauphine lives a Swiss clockmaker by the name of Maximilian Reinhart. Doctor Reinhart is said to be one of the finest clock-makers in all of Paris. An exceptionally gifted man. He seems respectable. He makes toys for the rich and gives alms to the poor. But some say he does strange things; that he's not what he seems.' He leant back. 'His maid has given notice to quit.'

A silence. Madeleine could hear high false laughter from the room above, the slamming of a bedstead against the next wall. Coraline came back into the room carrying a tray on which cups of chocolate clinked. She set it down on the table, then sat next to Madeleine, so close that she could smell the rosewater and sweat on her cleavage, the caramel on her breath. 'You've told her, then?'

'We're getting there, my love.'

'You wish to set me up as a spy,' Madeleine said.

'We need you to find out what he's up to,' Camille Dacier said. 'Establish what kind of man he is.'

Coraline put a cup of chocolate into Madeleine's hand. She hadn't whipped it properly. The cocoa had settled on the top.

'What kind of man d'you think he is, Monsieur?'

'I think him a very clever man, but an odd one. There are risks that must be checked.' He and Maman exchanged a look. 'There are rumours.'

'What do they say, these rumours?'

'That he engages in strange practices – that he carries out certain unnatural experiments.'

'What experiments?'

'That's for you to establish. But it's whispered he takes his work too far. That he'll do anything to achieve his aims. We need you to find out if that's true.'

Experiments, clockmakers, toys for the rich. What did she know of these things? 'Why d'you need to find out?'

'As you're aware, I work for Inspector Meunier. But it's not only him that I work for. A man must have several masters to survive in this city. In this case I answer to a very powerful person. A person who wishes to be sure of Reinhart's character and references before offering him an important role.'

Madeleine kept her gaze on her chocolate. 'What role would that be, then?'

He didn't answer that. 'We all like to check a person's references before inviting them to live in our house, don't we?'

She didn't reply.

'Not yours, though. No one will be checking your references in this case because I'll make sure you're recommended, your name whispered into his ear.' He leant forward again. 'Won't get another chance like this one, will you, a girl with your particular history?'

Madeleine looked at him, into his strangely coloured eyes. He was right, of course. She was damaged goods; a scarred, sullied *cocotte*. Twenty-three years old and still beholden to the bitch who'd given her birth; tied not by love nor duty, but a belief that – if turned out into it – the cauldron of Paris would swallow her up. For Paris only kept alive those who paid, and Madeleine, as she'd been told many times, had precious little to offer.

'We owe it to Monsieur Dacier to be accommodating,' Maman said, falsely sweet. 'He's been very good to us. Looked after us.' She stared at Madeleine, a smile warming her lips, but her eyes cold and grey as the Seine. 'You'll accept the position, my dear.'

Camille was watching this exchange with a faintly amused

14

look on his face. He was, Madeleine understood, a little man who enjoyed what power he had.

'What exactly would I be required to do, Monsieur?'

'You'll be a *chambrière*, a maid of all work, much as you are now. You'll also be *femme de chambre* to Reinhart's daughter, Véronique, just as you are to the other girls here. Only your mistress will be rather less ... worldly.' He curled his lips at Coraline. 'You'll note who visits the house. You'll listen in whenever you can. You'll record any unusual activity, any interesting conversations. You'll read any letters, journals or diaries that you can get your hands on. And, every week you'll report back to me.'

'On Doctor Reinhart.'

'On anything you see in that house. You'll have thirty days to complete your task, to determine what these experiments are that he is carrying out, and to establish whether he's a man to be trusted.'

A month away from Émile. She didn't like the thought of leaving him alone here, and she knew full well that he'd hate it.

'Won't the clockmaker know me, though, for a girl of low birth and upbringing? I've only ever worked here – I'm hardly a lady's maid. Why not ask someone who really is?'

'We'll train you up a little over the next fortnight – polish your accent, get a real *femme de chambre* to show you how it's done – but I suspect Reinhart will barely notice you. The rich don't generally consider their servants to be real people. And the daughter's as fresh as a daisy. She won't know any better.'

'How old is the daughter?'

'Seventeen years.'

Madeleine shifted in her seat. Not much older than Suzette when she died. 'What's she like, then?'

Camille picked at his teeth. 'Green, childish, just returned from the convent school she's been locked in for ten years. I

15

doubt you'll get anything of interest out of her, but it's a good route in, you see?'

She pondered this. An undemanding mistress, clean sheets, a warm house. A way out. Yet it was still there, that nagging feeling that the whole enterprise was a snare.

'What would I be paid?'

'I've dealt with that,' Maman said, quick as lightning.

Of course she had. Just as she dealt with what her girls received for their services, what went to her for their 'upkeep'. Well, Madeleine wasn't having that. Not if she was putting herself in danger, which she must be if they were willing to pay. She looked squarely at Camille. 'If I'm to do it, the money must come to me directly.' She could see Maman's face darkening, the fury rising off her like smoke. 'Then we divide it up.'

Camille laughed. 'Ah, she's her mother's daughter.' A pause as he assessed her. 'Very well. Five hundred livres direct to you once the job is complete. How you share it with your mother is your business.'

Five hundred livres! Maman had negotiated well. But this was too easy, too slick. 'And after I've completed this task, what do you intend for me then, Monsieur?'

'Why, then you'll rise in the ranks, Demoiselle Chastel. If you succeed, you'll be given further opportunities, more responsibility, more money.'

If she succeeded. What if she didn't? What if they found her out? Police *mouches* were hated, detested. Those who were outed were stoned in the streets and those who lived became outcasts. But then, really, what was she now? And what other opportunity would she have, a marked maid from a brothel? She'd live out her days here, watching as wave after wave of girls became riddled with pox; watching as Émile was corrupted or killed, by a cove or by his own weak health. She ran her tongue over her teeth.

16

'But if they do work out what I am – or if I can't find what you need – what happens to me then?'

He didn't answer that. 'If you're good enough,' Camille said, 'if you're as clever as Madame here has claimed, then you won't be discovered, and you'll find what I've asked.' He tilted his head to one side. 'D'you think you're smart enough, demoiselle, to fool an entire household? To steal their secrets and bring them back to me?'

Madeleine held his gaze. She'd been dissembling almost her whole life, one way or another. She could deal with a clock-maker, couldn't she?

Maman leant forward and put one of her hot hands on Madeleine's. 'This is it, *ma petite*. This is your opportunity to shine. Do this for us, will you, now? Make your old mother proud?'

Madeleine paused for a moment, her eyes on Maman's stretched smile, on her teeth, blackened by rot, by years of bonbons and sweetmeats. 'Yes, all right. I'll do it, since that's what you want. I'll do the best I can.'

But of course she wasn't doing it for her mother, nor Coraline. She was doing it for the money, for Émile: she was gambling it all for the chance that they might be able to escape.

2

Over the next fortnight the weather grew harsher, colder, the price of bread soared, birds fell frozen from the sky. Men lit fires in the streets to save the paupers from dying, children slept on lime kilns, newborns froze on orphanage steps. Snow fell thickly, silently as death. It shrouded the filth that covered the streets; it blanketed the putrid river of refuse that ran down the centre of the roads so that the alleyways of the slums shone like the salons of the rich. And when the day came for Madeleine to leave Maman's 'Academie' to travel to the clockmaker's, the ground was frozen to a glimmering sheet of ice. She dressed quickly and quietly in the winter dawn darkness, listening to the scuffling of the mice in the walls. Then she bent to kiss Émile's sleeping head, breathing in his little boy smell, praying (with words not usually used in prayer) that she wasn't – like one of those scuffling mice – walking straight into a trap.

Treading cautiously in her thin leather boots, she left the still-dark house and hurried down the Rue Thévenot, past the shuttered windows of the locksmiths, fan-makers, illicit book-sellers. Good riddance to the lot of them – she was off to better things. As she neared the doorway where the street girl slept, Madeleine put her hand in her pocket. The doorway, though, was an empty space and Madeleine herself felt suddenly hollow.

A cup remained, which she'd used for soup, but the girl herself was gone. Possible, of course, that she'd found some charity, but more likely she'd been driven out. Quite possible she was dead. Well, it was only what she'd expected, wasn't it? She must think no more of it now.

At the corner of the Rue du Bout du Monde, Madeleine caught the smell of coffee and there stood old Marie, a tin urn on her bent back, a pewter mug glinting in her hand. 'Stop for a minute, love? Two sous a cup.'

But Madeleine couldn't stop. What kind of maid, they'd wonder, gets up so late? What sort of mistress runs a house that lets her? She walked on, past the grand steps of the Louvre where the snow had been heaped into pyramids, and onto the Quai de l'Ecole where boats knocked against their moorings and the swollen corpse of a dog floated silently past. Then across the Pont Neuf where the twin clock dials of La Samaritaine showed her she had only a few minutes to go and where the snow had been churned with ash, mud and dung, carved through with cartwheels and printed by horses' hooves, and where mufflered clerks, tailors, vintners, rat catchers, barbers and bookbinders were making their way to offices, workshops and the homes of the *gens de qualité*. On the unsheltered expanse of the bridge, the wind was an icy blast stinging skin and blowing hat feathers. Madeleine glanced back across the grey water to the Right Bank, to the place that'd been both home and prison for the past eleven years, the place that'd half killed her inside. Perhaps this would be a chance to live again. Or perhaps it would finish her entirely.

Then she was on the Île de la Cité and at the tip of the elegant Place Dauphine with its triangle of tall houses: the homes of gem dealers, pearl traders, mirror makers and watchmakers, who'd only now be shaving and sipping their morning chocolate and unlocking their precious wares. Hurrying towards the far end of the square, she caught the scent of baking dough and saw the line of shivering servants waiting to buy the day's bread. She saw the hands of two beggars held out to those emerging with floury loaves and the answer of averted faces. In Paris, if you couldn't earn your bread, then your bones might be made of glass, for others seemed to look right through you.

As she finally reached the far end of the Place there began all around her a great ringing from the bells of all the churches on the island and of the city itself, carrying across the water, each bell with its own peal, chiming together in a vast rush of sound, as though time itself were chasing after her.

Then she was there on the steps of the house with the sign of the golden clock, where she scraped the snow and shit from her boots, raised her hand to the brass handle, took a deep breath and knocked.

For a minute or so she stood, bag clutched in her hand, staring up at the tall sandstone house, the long windows that gleamed coldly in the growing light, obscuring what was within. The building seemed to tilt forward so that it stared down at her in disdain, seeing her for what she really was. It was grand, certainly, compared to what she'd come from, but not the sort of grand she'd expected. From what Camille had said, Madeleine had imagined something plusher, fancier, a regular palace; but the house had instead the gaunt, forbidding look of a poorhouse or an asylum. Standing there, she was reminded that everything she knew about the place, everything she knew about the clockmaker, came from a man she wouldn't trust with her purse.

At last the door opened and a man stood there unsmiling. She'd never seen anyone so dark, least not close up. He was dressed in powder-blue livery with gold braiding. A valet. He said only, 'You're the new maid,' and stood back to let her in.

The hallway was of chequered marble, black and white, and all was silent save for tick, tick, ticking. Along the hall was a line of clocks: a long case clock with ornate brass work, a gilt clock swarming with tiny gold figures, a pedestal clock with a silver bird that opened and closed its beak. Of course there was nothing so unusual about the clocks – this was a clockmaker's house after all – but the motion of the machines, the constant ticking, made her doubly nervous. For they didn't all beat at the same time, but were out of sync, like panicked, uneven heartbeats.

The valet walked before her along the hallway, not speaking, so that she was free to cast her eyes along the walls, at the pictures that hung above the clocks. Not that they were pictures exactly, but sketches of human bones, and then a man drawn stripped of skin, the sockets of his eyes empty. There was a cold smell to the place, of wax polish and lilies, nothing like the all-too-human reek of sweat, powder, tallow-smoke and piss that she'd lived with at her mother's house. She'd a sense, though, that there were people here, breathing quietly; watching. She followed him down some steps to the kitchen, a large, tiled room hung with copper pans, cured hams, a bright brace of pheasants. A woman in a white cap stood at the stove, stirring coffee in a pan. Her back seemed brittle and thin.

The woman turned. 'About time.'

It was, what, five minutes after eight? Madeleine gave a brief curtsey. 'The roads were terrible icy, Madame.' *And given how risky this whole rig is, you're lucky I'm here at all.* She already had the sense that she shouldn't have come; that something in this house was askew.

She kept her eyes to the floor but knew that the woman was staring at her, wondering what was wrong with her face. 'Joseph, you can take her bag to her room.'

The man nodded, his expression still a mask. He picked up her case and left.

The woman took the pan from the stove, set it down and walked over to Madeleine. She had a dull sort of countenance: small features, muddy eyes, skin the colour of bacon rind. She was a few years older than Madeleine: eight and twenty, maybe more, years that had sucked her of joy. 'You're Madeleine, then. My name's Agathe. I've worked here for five years, but that's now at an end.' A pause. 'Your duties here are many, but you'll manage if you organise yourself. You were at a cloth merchant's before, is that right?'

'Yes. On the Rue Saint Antoine.' Madeleine avoided her eye.

'You were a maid of all work?'

Madeleine nodded. All types of work were covered, certainly: from virginal and innocent through exotic and bizarre, to the sadists and whipping culls.

'Well, here there's a cook, Edme, and there's Joseph who you've seen; he looks after Doctor Reinhart. We had a kitchen girl for a short time, but she was worse than useless. You're to help Véronique, the daughter, with dressing and her toilette. Until a fortnight ago the girl was living at a convent, but she's seventeen now and must be helped to behave and look as she should, what with no mother around.'

'What happened to the mother, then?'

'Dead in childbed, when the girl was born.'

Common enough. Common as sparrows. She would think no more about it.

'And Doctor Reinhart – he's ... a good master?'

The maid narrowed her eyes. 'He's fair enough, yes. You'll find his manner a bit strange, maybe.'

'In what way?'

'You'll see when you meet him. He's off visiting a customer at present.'

'He sells clocks?'

'Clocks, and other mechanical objects. He makes metal creatures that move.' The maid twisted up her face. 'Queer things. Wouldn't want them in my own house, if I had one, but rich folk seem to like them.'

Madeleine thought of the heartbeat ticks of the clocks, the bird with its moving beak.

'And he's training her to make them herself,' Agathe added quietly.

'The daughter?'

'Yes.' The maid raised a sparse eyebrow in disapproval. She walked to the table and picked up a coffee pot.

'And you, Miss Agathe?' Madeleine asked. 'Are you staying for—?'

'I'm leaving today.'

'For another position?'

'I need to go to my family.' She pursed her lips. 'My mother's ill.'

'I'm sorry to hear it.'

The woman nodded. Perhaps she'd thought Madeleine's sympathy genuine, for when she looked up her eyes had lost their dull glaze. 'There are comings and goings here sometimes. At night.'

Here was something. 'What d'you mean? Who comes and goes?'

'My advice to you is to keep your questions to yourself. Easier for everyone that way.'

Madeleine would've asked more, but she heard footsteps on the kitchen stairs, the swish of fabric on the floor.

A girl ran into the room: slight, sharp-featured, fair. She wore a crumpled green peignoir with brocade slippers and looked, Madeleine thought, like some elfin being from a fairy tale, not a girl from the world she knew. She stopped when she saw Madeleine, and stared at her, unblinking. An uneasy

23

silence hung in the air until Agathe said, 'Véronique, this is Madeleine, who's come to assist you.'

The girl's eyes were a strange colour: light green, veined with amber.

'Good morning,' she said at length. 'I hadn't expected … That is, I'm pleased you've come.'

She did not sound pleased, though, for her voice was flat. She was beautiful, Madeleine realised, in an other-worldly sort of way, her skin smooth and creamy as a bolt of ivory silk. It would be difficult not to hate her, but perhaps that was just as well.

'I've been instructing Madeleine on her duties here,' Agathe said, 'but you'll have to tell her how you like things to be done.'

'Well, there's not a great deal to tell. I've never had my own maid before. I've always managed by myself.' Véronique looked at Agathe. 'Where's Edme? Has she not yet made breakfast? I confess I'm starving.'

Starving. It wasn't a word you used if you'd ever known real hunger, or if you knew of the growing flood arriving from the countryside, barefoot and hollow-eyed, desperate for blackened bread. But of course the girl had been sheltered in a convent, in this house, this life. So much the better, Madeleine thought. The greener the girl, the easier her task.

'Edme's fetching fresh rolls,' Agathe said. 'And I'm making your coffee. Madeleine, you go with Miss Véronique and help her with her toilette. I'll bring up the breakfast shortly.'

The girl walked before Madeleine, up the kitchen stairs and through a room lined with shelves of glass jars filled with nightmares – a two-headed piglet, twin snouts poking upwards, and a black-striped snake, cloudy eyes open, twisted in on itself.

Véronique glanced behind her. 'Preparations,' she explained.

'My father trained as an anatomist. You know what that is?'

'I think so, Miss.' Madeleine knew, all right. It meant men who cut things up, who ottomised corpses; whose hands were coated in gore. 'But I thought he was a clockmaker.'

'He is, but he makes much more than clocks. Come this way.'

As they walked, Madeleine glimpsed a jar containing what seemed to be an unborn child, its skin like a poached egg, its eyes squeezed shut, downy hair on its malformed head. Was this the result of the experiments Camille had talked about, or the work of some other man? What kind of person would keep a baby pickled like an onion in a pot?

Véronique opened a door to reveal what must be the doctor's workroom, for it was full of lathes, a machine with a large brass disc, some sort of engine, shelves and cupboards stacked with boxes and books. Next to the long window stood a high workbench cluttered with callipers, hacksaws, odd wood-handled tools. Further implements hung from the walls – for clockmaking, she supposed, but some had the nasty look of instruments used for torture. The whole place had a distinctive, unfamiliar smell: part soot, part chemical, part mystery.

'This is where he creates.' Véronique opened a drawer beneath the workbench and removed a small wooden box. Inside was a silver spider that Madeleine took at first for a brooch, but before she knew what was happening, it was running across the table towards her with the hideous furtive motion of the real creature, thin silver legs rasping on the wooden surface, and it was all she could do not to scream. Abruptly, when it was only a few inches away from her, the thing came to a standstill.

Véronique smiled but her eyes remained watchful. 'Wonderful, isn't it? But it isn't yet finished.'

Madeleine merely stared, her heart kicking in her chest, the taste of bile in her mouth. She wasn't going to be taken for a fool by a girl of barely seventeen years. Looking at the spider closely, Madeleine saw that its body was made of two pieces, like a box, and that it was cleverly worked so that it looked almost unbearably real. This, then, was what Agathe had meant by 'creatures that move': machines like the clockwork figures she'd seen at fairs, only tiny, beautiful, horrible. 'Very good, Mademoiselle,' she said blandly. 'Now we should get you dressed.'

Véronique's boudoir was painted a dark petrol blue that made Madeleine feel as though she were sinking beneath the sea. On a fire screen painted birds flew, and in a corner glinted a wardrobe made of ebony, tortoiseshell, bronze. From a heavy gilt frame the oval face of a woman looked down at them and, beneath it, a curious worn doll stared straight ahead with its empty green glass eyes. It was like no room Madeleine had ever seen, but she kept her expression as blank as the doll's.

Véronique settled into the chair at her dressing table.

'Shall I help with your hair, Miss?'

'Yes, I suppose you should.'

Madeleine took up the silver brush from the table, feeling the heft of it in her roughened palm. Véronique's hair was thicker than Suzette's had been, fairer, and cleaner, for you could only keep your hair so clean when having to wash it beneath a stone-cold pump. She met the girl's eyes in the mirror. Véronique had been studying her face.

'So you are come to look after me. How old are you?'

'Nearly four and twenty.'

Véronique gave a half-smile. 'Then perhaps I'll be like your little sister. Do you have sisters?'

Madeleine tried to keep the brushstrokes smooth. There was something about the way the girl spoke, as though she was mocking her. 'I have one sister, Miss.' *And another not long in the ground who's taken a piece of me with her.*

'Younger or older?'

'Two years older.'

'Is she pretty?'

She thought of Coraline with her hazel curls and practised smile, Suzette with her heart-shaped face. *Though God chose to give me girls, at least he made them handsome.* 'Yes, very pretty.'

'Is she a servant also?'

'No, she's an actress,' she said, and in a way she was: faking desire night and day to awaken desire in others.

'Where?'

'Here and there. Nowhere grand. Probably not the places you've been to.'

'I've never been to the theatre. I returned here only recently and it's not something my father does.'

Madeleine didn't know what to say to this. Why live in Paris and not go to the theatre, the ballet, the opera, if you were as rum as they? She knew what it was to be penniless in Paris, to have your face pressed up to the shows and the shop-fronts, the glittering windows and the rooms full of laughter, knowing you could never go in. If the doors were open to you, why not enter them? Something here was amiss.

'Where are your curling tongs kept, Miss?'

'I don't have any. Perhaps I should.' She was still staring at Madeleine, doubtless working out why Madeleine wasn't herself an actress, and how it might feel to be the scarred sister of a very pretty girl. Worthless, she might have told her, and sometimes glad, and sometimes sick with guilt.

Madeleine took up the silk petticoat that was lying on the bed, and Véronique removed her nightgown, exposing her milk-white skin. She wore no corset, no stays. 'Raise your arms then, Miss.' Madeleine dropped the shift over her head. 'Have you chosen a dress?'

From her armoire Véronique selected a mauve gown, plain but of better quality than anything Madeleine's sisters had ever worn, worth fifty livres at least. She fastened the tiny bone buttons for Véronique, tightened the sleeves, and began to lace the bodice.

'You were a *femme de chambre* before, then?'

Madeleine did not like this questioning. She continued with her lacing, thinking of the training she'd received in the past two weeks from a maid Camille had bribed. *You must only speak when they question you,* the girl had told her. *And keep it as dull as ditchwater. They don't really care what you think.* 'Yes, Miss. For many years.' Ever since her own value had plummeted.

'You don't mind?'

'It's better than many jobs.' *But yes, of course I mind. To be the least valued member of the household, dressed in the outcasts of others? To be the one who has to empty the reeking chamber pots because I'm only half-price as a whore?* She had a path out now, though, minded it was a narrow one. If she tripped she'd a long way to fall.

There came the sound of an outside door closing.

'Your father?' Madeleine's voice came out too sharp, her accent too hard. She'd been doing her best to soften it.

'No, he said he'd be away until tonight. It must be Edme.' A pause and then she said quietly, 'You mustn't be afeared of my father when you meet him, by the way; mustn't mind his manner. He's sometimes better, I think, with machines than he is with people.'

Machines. Like the scuttling spider, the silver bird. 'Oh, I doubt I'll be frighted by him,' Madeleine said, folding Véronique's nightgown and placing it beneath her pillow, and thinking of the men she could tell her of: men with balled fists

28

and red eyes, or – worse – men who were cold and cruel. But then, that was precisely why Camile had chosen her, wasn't it: her knowledge of such men, her ability to bear it? She knew then that the clockmaker might not be simply strange. He might well be something much worse.

★

Edme, the cook, was a heavy-jawed woman with grey-streaked hair plaited into a thick braid. She regarded Madeleine with some suspicion from beneath her black brows, looking her up and down as though considering an item offered for sale to work out if it were stolen. 'How many years you been in service, then?'

'Nine, Madame. Since I was fourteen.'

'Have you served at table before?'

'I have.' Served pies and chops from the pastry shop, and pitchers of cheap wine to men too drunk to question the quality. 'You needn't worry on that account.'

The woman frowned. 'We're very particular about how things are done here. Don't go thinking you know everything. There are rooms that must be kept locked and things that must be done a certain way. You must always check with me.'

'Yes, Madame.' And a madam she was. 'I'll do my best to learn.'

Edme gave a curt nod. 'D'you take coffee in the morning or chocolate?'

'Coffee, if you please. With milk.'

'Very well. I'll bring you water for washing at quarter past five every day. You must be up and lighting the fires by half past.'

Madeleine felt fagged at just the thought of it. 'Of course, Madame.'

'And there is no waste here, nothing thrown away. Any food that's left over we collect and give to the poor that come to the door. That's the master's orders.'

Peculiar. 'He's a kind man, then, is he, Madame?'

'Fair, is closer to it. Believes all humans to be equal.'

That was odd, Madeleine thought, especially for an *haute bourgeois,* usually so obsessed with status.

'And none of this "Madame" business. You'll call me Edme. We use first names here. Except for the master, of course. He's "Monsieur" or "Doctor Reinhart".' She gestured to the table. 'There's bread here for breakfast, if you haven't yet eaten. Not as fresh as it might be, as the baker's lad's made off, it seems.'

Something tugged at Madeleine's mind like a hook. 'Made off where?'

'Who knows? The baker found him gone this morning, and him usually such a good boy. Jam too,' she said, pointing. 'Good jams. I make all my own. Only way you can be sure what's in them.'

Madeleine eyed the line of pots of glistening jam: the purple of damson, the red of strawberry, all with tiny silver spoons resting on the top. 'I've never seen the like.' That at least was true. There was no jam-making went on in Maman's house, though Lord knows there were plenty of plums.

The cook gave a grudging nod, but her face had flushed with pleasure. She pulled out a chair for Madeleine. 'Go on – eat. You'll need your strength for there's much work to be done in this house and not necessarily that which you're used to.' She paused. 'You came well recommended, I hear.'

Madeleine swallowed. 'I'm a hard worker.'

'Well, we'll see, won't we? We'll see how quickly you adapt.' The way the mort was looking at her, Madeleine feared she'd seen her written references for what they were: lies in black lettering. But then maybe she looked at everyone like that, weighing them for their worth, assessing them for hidden flaws in much the same way as Maman did with her girls. At any rate, she'd have to be careful with this one. Always on her guard.

'Sit. Eat. Drink your coffee. Then Agathe will show you what needs to be done.'

After breakfast, Agathe took Madeleine about the house, opening the doors to the parlour, the workroom, the bedrooms, the dining room, explaining the chores to be carried out in each: how the weekly wash was to be done, the parquet polished and the silver cleaned, the cupboards scrubbed out with vinegar, the floors spread with lye then sand, the boards scoured until they were spotless. Though Agathe showed her everything she must do about the place, Madeleine had a sense of things being kept back, of some secret just out of sight. Every room they entered was richly furnished with heavy velvets and silks of burgundy, chocolate, forest green, the windows hung with damask drapes, so that the house seemed to suck in most light and sound, swallowing footsteps and what she thought might be voices, leaving only the ceaseless ticking. Everywhere she went the clocks appeared – on shelves, on cupboards, on mantelpieces, looming at her from hallway walls.

In a room to the front of the house, Agathe unlocked a mahogany cupboard, saying, 'This is where he keeps the finished ones, for showing to potential customers.' She opened the cupboard doors and stood back, arms folded, nodding to Madeleine to take a look. And there, staring out of the shadows sat the most curious creatures Madeleine had ever seen: a bronze mouse studded all over with hundreds of tiny pearls, an enamel owl with eyes of agate and feathers of silver and gold, a turtle made with a real turtle shell, gilded Neptune sitting astride it. Sensing motion, Madeleine looked to see a silver bat with leather wings, hanging from the very top shelf.

'I thought it best to show you them now, so you don't get afrighted,' Agathe said. 'They move when he winds them, you see. And they've a nasty habit of making a sudden movement on their own. It's Joseph who usually keeps the keys. Myself, I keep well away.'

Madeleine stared at the mouse, at its blind red eyes and fine golden whiskers. Valuable, certainly, but queer. 'What are they, though? How do they work? How does he make them move?'

Agathe shrugged. 'They move as the clocks do, with screws and springs and wires and such, but I can't say I understand how he makes them. Seems much like magic to the likes of me. I'll shut them back away.'

As the doors closed, Madeleine took one last look at the mechanical bat, at the thin black leather stretched skin-like over silver bones; imagining the hands that made it. What kind of man would take such care to make such uncanny things?

'This is the master clock,' Agathe informed Madeleine, as they left the room and passed an immense walnut clock in the hallway. 'The clock against which all the others must keep time. It's Joseph's job to wind them, but you must keep an ear out for whether they're striking correctly. It's one of your most important tasks in this house: to ensure that nothing slips.'

Madeleine nodded, but her nervousness had deepened to fear, her stomach tightened to a knot. She'd never been so aware of each passing moment, of how little time she had to succeed. Thirty days, he'd said, and she knew if she failed, if she was discovered, there'd be no second chances, no reprieve. The police would hang her out to dry.

Lastly, Agathe showed Madeleine her room at the very top of the house.

'Here we are.' She opened the door to a small cell with

a palliasse covered in a brown blanket and a small charcoal brazier, unlit. 'It did well enough for me.'

'Yes,' Madeleine said, trying not to show her disappointment. Despite the luxury of the house, with its gilt furniture and white lace, this tiny room was little better than her room at home, and it had an odd smell to it that perhaps was Agathe's own. 'Thank you for showing me everything. I hope your journey isn't too long.'

Agathe stood for a moment looking about the room, as though wanting to fix it all in her mind. Madeleine tried to see it as she did, but couldn't. It was a meagre, musky place. The only other items of furniture were a plain wooden table and chair, and a tripod stand with a glazed tin bowl. So very little in a house of so much.

'Well,' Agathe said. 'I must go now. I wish you luck.'

'Thank you,' Madeleine replied, though luck seemed to her an odd thing for Agathe to wish on her. Why did she think she'd need luck, exactly, in this house? She wanted to ask her more: about her master, about Véronique, about the night-time arrivals, about her feeling of being watched. But Agathe, she realised, had tears on her cheeks, and was turning her face away.

When Agathe had gone, Madeleine lit the stump of candle on the table, unlocked her case and took out her second dress, her nightdress, her shift, petticoats and stockings. Everything else – the paper, the quills, the pot of ink – she left in her bag, which she locked and stowed beneath the bed. Servants weren't supposed to be able to write, which, she thought – trying to reassure herself – must be partly why she'd been recruited. Oh for her father to see her now, to see how his education had paid off. He himself had been unable to write properly at the end, his hands so often seized with tremors. In the last few months, it'd been Madeleine who'd written his bills and letters and so had understood, even at the age of ten, that the life she knew was sinking.

'You will write to me at least once a week, without fail,' Camille had instructed her before she left for the clockmaker's house. 'You will tell me everything you find. You will take your letters to the Café Procope and if I need you I'll send a message.' *But what if I need you*, she should have asked. What if something goes badly wrong? She thought of Émile, who would now be lying in bed, for the first time in his life on his own. She should've asked more questions, she should've done her own research, before agreeing to come to this house.

As Madeleine stood up, she heard the front door opening, then male voices. She pushed her bedroom door ajar to hear better, but the conversation had stopped. Madeleine made out footsteps on the stairs to the other side of the house, then another door closing and she felt her heart race, knowing that it must be him, returned at last: Doctor Maximilian Reinhart, Master Clockmaker of the Place Dauphine, performer of un-natural experiments.

Madeleine moved to her window. It looked down onto the square, now lit with flickering lamps. Opening the window, she could see the stone lions and eagles of the Palais de Justice; the narrow, uneven houses that crowded the Left Bank. She could hear the plash of oars as people were rowed away from the Île de la Cité, shouts from revellers on the other side of the Seine.

'By the fourth week,' he'd said, 'you'll know what these experiments he conducts are, and furthermore you'll have formed a conclusion as to whether the man is merely odd or of different order of strangeness.'

It was a clear warning sign, she saw that now. They'd come to her not just because she could read and write, but because they deemed her expendable. And they knew that whatever happened, her mother would keep her mouth shut.

Only Madeleine wasn't so easily crushable. She hadn't survived this long to be ground down, like the seeds in the apothecary's mortar. She'd succeed because she had to. Her sparkle might have dulled, but Émile was a gem, and she'd promised to get him out.

3

On the second day she saw him. 'It's best I present you to my father,' Véronique said, as though Madeleine were a new piece of silverware that he might want to inspect for dents. 'He'll want to know that you've arrived.'

Doctor Reinhart stood at the large table in the centre of the room, so occupied with the work before him that he didn't glance up as they entered, and Madeleine had time to take in the tall frame, the silvery black hair, the magnified layered spectacles that made him look like a giant fly. Then she saw what he was working on and realised it wasn't, as she'd assumed, some piece of metal, but a rabbit, flayed, its skin and fur cut back, its guts exposed. The man looked up. He wore gloves and held a scalpel in his hand. Madeleine gave a curtsey, which seemed hopelessly out of place.

'Father,' Véronique said hesitantly, 'this is Madeleine. The *femme de chambre* you employed for me.'

Reinhart lifted up the outer lenses of his spectacles so that he could survey her, looking very closely at her face. 'Yes, so she is. A distinctive injury.'

Madeleine felt the blood flood to her cheeks. Why had he been told of her scar, and why had he still employed her?

The clockmaker had returned to slicing up the rabbit. There was a strange smell to the room, of flesh, medicine, vinegar. 'Why a rabbit, you are wondering.'

Madeleine wasn't sure whether he was talking to her or Véronique or both of them.

'A silver rabbit,' he continued. 'She wants me to make a silver rabbit.'

'Who, Father?'

'The Maréchale de Mirepoix. There is some forest-themed ball at Versailles. She wants an automaton with which to impress the guests. She wants the thing to hop. Ha!'

Automaton. It was a hard-edged word, like cut glass; like herself. Not a word Madeleine had heard before. She scanned the room, taking in the line of cupboards, one open, revealing dozens of glass vials, a pile of books, a miniature model of the human form.

Doctor Reinhart peered at her through his spectacles. 'I need you to get me a different one. A younger one. Alive.'

'I beg your pardon, Monsieur?'

'A live rabbit. Go to the market, see if you can find any there. If you can't, find some boy to trap one for me.' His voice was like his scalpel: sharp and precise.

Madeleine glanced at Véronique, but she didn't seem to see anything unusual in this. She had entered a house of lunatics.

Then came a strange rolling noise, like a marble running along the floor, followed by a clicking, then a chiming, and a great swelling of sound as the many clocks in the workroom and the rooms beyond began to strike eight o'clock. Reinhart and his daughter stood motionless, waiting, until the cacophony had stopped, and then Reinhart continued, 'While you're there, see if you can get any ox tongues.'

'For ... your work?' she asked.

'No, girl. For dinner. I've a yearning for ox tongue. Edme

37

will tell you what else we need. Véronique, stay here. I need your eyes. Put some of those gloves there on.'

The girl obeyed mutely, but her movements were unsure. Madeleine could've sworn she was nervous.

It was a blessed relief to escape the house, with its non-stop ticking and its unpredictable inhabitants. There'd been something deeply strange about Reinhart, Madeleine thought; something not quite human, or at least not like the humans she knew. Then again, he was a scientist, a man of learning, and she'd met precious few of those.

Carrying a wicker basket, Madeleine headed on foot through the Place Dauphine where barbers were leaving the grand houses, wigs in one hand, cases in the other, their clothes covered in flour. She walked quickly northwards along the Pont Neuf where the stallholders were setting out their wares, then onto the Rue de la Monnaye and followed a series of winding and blackened streets, where the distinctive reek of the Cimètiere des Innocents was eventually replaced by the smell of charcoal and day-old fish; she had reached Les Halles, the city's most notorious market.

It was the wrong time to have come – every servant and shopper in Paris seemed to be there, swarming over the filthy flagstones and around the pyramids of eggs and crates of oranges, skidding over cabbage leaves and fish guts and damp straw in their hurry to find the best *bonne affaire*. Among them walked vendors carrying casks of sour wine, crates of old shoes, platters of shrimps, rat bane, baskets of spices. Scrawny, rouged street girls, scant breasts pushed up in borrowed corsets, rubbed up against passing men to let them know what else was for sale. A pack of dogs ran amok, beaming. Madeleine selected the ox tongue from the meat slabs, being careful to check it was properly fresh, not just painted to make it seem succulent.

Then she made her way over to the live produce tables, where geese hissed from their pens, chickens pecked at the ground, and cages of ducks were piled up high. Was there a crate of rabbits, though? Not a bloody one. She was sure there'd been rabbits here before – a whole cage of them, squirming – but today was not her day. She wandered away to where she saw a woman wearing a hideous patchwork cloak pieced together from pieces of fur and fabric. In her hand she held a brace of hares.

'Excuse me, Madame. D'you know where near here I can buy a live rabbit?'

The woman looked at Madeleine closely. 'Alive.'

'Yes.'

'For a pet?'

'I think so.' She didn't think so. She suspected that the un-natural experiments Camille had spoken of were performed on living flesh. 'My master asked me to fetch one.'

The woman nodded. 'They'll be having the oddest pets in their houses, these nobles. Did you know the King's mistress, La Pompadour, has a pet monkey? It sits on her shoulder and eats her food.'

'Does it, now? But the rabbits . . .?'

'Well, there was a boy that sold rabbits over there,' she pointed, 'caught them himself and brought them in. But he's not been here at least a fortnight.'

'Has he moved to a different market?'

The woman shrugged, disinterested. 'No idea. There's a man on le Rue Denis, though. He'll have them. Who is he then, your master?'

'Doctor Reinhart his name is. He's a clockmaker.'

'Oh, I know what he is.' The woman's face had taken on an intent, malevolent look.

'What d'you mean?'

'Do you really not know?'

'I started there only yesterday.'

'And already he has you at his devil's work.' The woman pointed at Madeleine. 'They say he creates things that no man can create. That he's a magician.'

'No, a *mecanicien*,' Madeleine said uncertainly. 'A clockmaker.'

The woman pointed a gnarled finger at her. 'Those things he makes, whatever it is that he calls them — look at them closely. They're impossible things, made with dark magic.'

Madeleine moved closer to the woman, unease snaking coldly through her. 'Who told you that?' she said quietly. 'What else did they say? Why did they think it was black magic?'

'Plenty of people say it. How else d'you explain how he does what he does? How else d'you bring a creature to life when it's made of screws and metal?'

Madeleine felt her body relax. The woman wasn't a useful source — she was a half-witted crone from the countryside. 'He makes them with gears and cogs and such,' she said flatly. 'Just like he makes the clocks and watches.'

The woman shook her head. 'If you knew what was good for you, you'd leave that house now. You'd get yourself out of harm's way.'

Ah, but she'd been in harm's way for many a year, and this might be her way out of it. 'Well, thank you, Madame, for all your advice, but I reckon I'll take my chances.'

Madeleine left Les Halles and headed for the Quai de la Mégisserie, to the shop that had once been her father's. She knew as she entered, as soon as she inhaled, that it'd been a mistake to come here. Until his final months, her father had taken great pride in the shop, regularly retouching the paintwork and keeping it clean and well stocked, with parrots and cockatoos, squirrels and lizards, green monkeys, miniature goats and rabbits that she and Suzette had been allowed to pet. Now, the place was dusty and fetid and mostly empty, save for mice in cages and puppies in boxes: sad little creatures who whined for their mothers and pissed on their furry feet.

The shopkeeper was a defeated-looking man with a yellowed bob wig and rheumy eyes. She didn't tell him the shop had once been her father's. What would have been the point?

'Do you have any rabbits, Monsieur?' she asked instead.

'Rabbits? Why, yes, I do.' He went into a back room and returned a minute or so later carrying a box, from which he lifted a small white rabbit with shining red eyes. 'Perhaps you'd like to have a look at him?'

Madeleine took the rabbit from the man and it was the silkiest thing she'd ever touched. She could feel its heart pulsing quickly beneath her palm and for a whole moment she wanted to take it to her little sister, for Suzette had always loved rabbits, had kept one for a pet in her room until the neighbour's dog had got at it. But Suzette of course was dead and gone, so had little use for rabbits.

'Thank you, Monsieur. He'll do nicely.' She put the rabbit back in the box.

When she returned to the house, Doctor Reinhart was sitting at a bureau by the window in his workshop, writing in a leather-bound book. To his side was an astonishing eagle with talons of brightest gold. She carried the box over to him and set it down, nervy that at any moment the eagle would spring to life. Reinhart glanced at Madeleine, then at the box. As she opened it, he smiled.

'Ah. Our template.' He held up the rabbit before him and looked closely at its twitching nose, its trembling whiskers. 'Very good.' Then, realising Madeleine was still standing there: 'That will be all, girl. You may go.'

Madeleine hesitated. 'Are you ... are you going experiment on him?'

Still holding the rabbit, Reinhart turned to look at her. 'To understand life, Madeleine, we must sometimes have recourse to death. The two are bound together; inextricable. You will come to understand.'

During the hours that followed, as she ran between her various tasks, Madeleine tried to get a glimpse of what Reinhart was doing, but the workshop door remained firmly closed. It wasn't until three o'clock that she was able to return, with the coffee and almond cake Edme had baked that morning. She carried in the tray cautiously, quietly, expecting to find the rabbit stretched out as the other one had been, its snowy fur peeled back. She prayed that it was at least dead. Doctor Reinhart was squatting on the floor, a notebook in his hand. He barely registered her presence, his attention focused on something ahead of him, out of her sight. As she put down the tray, she saw that the rabbit was on the floor, still alive and seemingly unharmed, chewing at the embroidered rug.

'We need something to feed him,' Reinhart said abruptly. 'Some vegetables. Some oats. Can you get them?'

'Well, yes, Monsieur.'

'And a house of some kind. A hutch. See what you can find.'

'You're keeping it?'

'Him. It's a male rabbit. I have decided to name him Franz.'

It was Joseph, the valet, who found a home for the rabbit by constructing a cage with some wire mesh bent over a box. She watched from the window as he worked out in the yard with the confidence and speed of someone used to making such things. Madeleine had spoken little to Joseph since her arrival at the clockmaker's house. It was, after all, never wise to get too close, and she found him vaguely unsettling. She wanted, however, to know what he knew; to understand what Reinhart was about.

She walked into the yard, the washing on her arm. 'Has he kept rabbits before, then?' she asked after a minute or so.

Joseph glanced up at her. 'Not rabbits, no. Geese.' He continued hammering.

'For eggs?'

'Golden eggs. He made a goose that laid a golden egg. Last year.'

Madeleine smiled. She'd been determined to stay aloof, but she couldn't help herself. And he smiled back, grooves appearing in his cheeks.

'I wish I'd seen that.'

'Yes, it was a fine thing. Three geese we had to get, in order to find the right one; several bats were netted for the bat he made. It will be the same with the rabbit.' He continued with his hammering.

'And he experimented on them, did he?'

'Well, he observed them, then cut them up. It's how he sees how they work. Then he pickles the parts he wants to keep.'

Madeleine thought of the delicate bat skeleton, the eyes she'd seen floating in a jar. 'Right.'

'He says,' Joseph continued, 'that he has to learn how the animal moves in order to reproduce it.'

'As' – she tested the word – 'an automaton?'

'Yes, that's right. That's what he calls them. Machines that move by themselves.'

'How do they move, though?' She thought of the woman in the rabbit-skin cloak and her claims of dark magic.

'With screws and chains and pulleys. Doctor Reinhart, he's a very clever man.' Joseph stood up to survey his work. 'There. That will do, I think.'

'It does very well. You have quite a skill.'

43

'The slave who builds cages.'

Her smile fell. This was what happened when she tried to make conversation. 'I didn't think you ...'

He shrugged. 'I'm free here, as much as any servant is ever free. But by law I'm still a slave, to be bought or sold.' He picked up his tools and walked away, leaving Madeleine to stare at the wire mesh, wondering if there was anything in the whole of Paris that couldn't be bought for a fee.

*

By the evening Madeleine was exhausted, drained by the strangeness of it all, the constant chores, the strain of playing a part. She'd have to force herself to stay awake, mind. She needed to get into that workroom. Madeleine waited until everyone was in bed, the house in darkness; until she could hear the soft sound of Edme's snoring from the next room. She waited until the house creaked with the sounds of the night. Then she took her nub of candle and crept downstairs, her bare feet almost noiseless on the polished floor, and hesitated outside the workroom door. After a moment, she pushed down the handle. It was open.

Inside, the room smelt of the ox-gall used to clean the floors, the animal reek masked by alcohol. The only sound was the slow ticking of the large table-clock. Keeping her hand as still as she could, she used the candle to light a rush lamp, which she carried with her as she went from cupboard to cupboard, illuminating the dark alcoves within. Her breathing was shallow, her heartbeat quick, as the flame picked out the oddities on the shelves: a small skull, tiny bottles of unknown solutions, a tray of coloured glass eyes. On opening a drawer she found it full of varnished human bones.

Could any of these be evidence of his strange experiments,

or was it the sort of thing every anatomist kept, eyeballs stuffed in their cupboards and drawers? How on earth was she to know what would count as 'unnatural'? To her none of these things looked normal.

She opened the bureau and rifled through it for Reinhart's letters and notebooks. One by one, she took out the papers and peered at them, trying to find something that might give a clue as to the nature of the man, something that might interest Camille. There was a bill from a goldsmith, other invoices from tradesmen, a list of what seemed to be tools, a diagram of a piece of machinery, a drawing of some kind of cage. Nothing that seemed important. Then again, she didn't know exactly what she was looking for. Quick as she could, she returned the invoices and moved on to the bureau drawers. There she found a letter from someone called La Mettrie, telling Reinhart about his discoveries in Leiden, about his theories on animals and souls with strange pictures in the margins of which she couldn't make head nor tail. She searched for a few minutes more but found nothing of obvious value. If Reinhart did carry out strange experiments, he didn't keep his writings on them here.

Before she left, Madeleine walked to the far end of the room, beyond the worktable and benches and towards the cabinets at the back. In the wavering semi-darkness she could see only the silvery edges to objects, but she made out the outline of a long rectangular box, the onyx gleam of black paint.

If she'd had more time she might have struggled to make herself open it. As it was – worried that someone might come upon her at any moment – she prised open the lid and let it rest on the wall behind. There, in the glimmer of the candle flame, she saw a girl, her dark hair laid out across her shoulders, her face white as bone, eyes open. Dead.

Madeleine felt panic rise in her throat like bitter juice. 'He may be dangerous,' Camille had warned, but, dear God, she hadn't expected this. So young the girl was, and beautiful; seemingly unmarked. There was no smell of death, no tang of blood. She was like a saint embalmed in a glass coffin, preserved somehow from decay. The horror pulsed through Madeleine's veins, but she must keep her wits together. As Madeleine brought the lamp closer, the flame flickered, lighting the girl's face in such a way that she knew at once that there was something wrong with her skin. It was not skin.

Madeleine blinked and the scene seemed to rearrange itself. What she saw now was a doll.

Madeleine brought the lamp still closer and leant over the model. The eyes glinted in the candlelight: glass. The skin shone as only wax could do. And yet, as she moved the lamp up and down, she saw that this was no simple doll. Nothing like the crude waxworks she'd seen in travelling shows. She was perfect, horrible. Every feature, every contour was precise and true. Peering closely, Madeleine saw the delicate veins painted on her chest, the pink of her nipple. She shuddered. The whole of her was too real, too close to the truth. The candlelight rippled and it was almost as if she moved. With a sudden movement Madeleine shut the box.

Trying to calm her breathing, Madeleine returned everything to its original position, relit her stump of candle, snuffed out the rushlight and quietly closed the door behind her, feeling a surge of relief tinged with unease. She'd found nothing to suggest perverted experiments, nothing obviously wrong, nothing to confirm what sort of man Reinhart was. Yet she felt, just as she had when Agathe had shown her about the house, that something was lurking, just out of sight. As she emerged into the hallway, she had a strong sensation that someone was watching her, but it was only the clocks, their blank faces seeming to observe her as she made her way past them and back up the stairs to her room.

4

Véronique

Faster, go faster, there isn't time. Véronique ran through the cloisters, the stone arcing above her, the night air dark and chill, knowing that she must reach the crypt before the bell sounded. But her body refused to move quickly enough – pushing against the air as though it were water – and she wanted to shout that she was coming, was coming as fast as she could, only her mouth was a gaping hollow that refused to make a sound, and as she pushed open the cell door there came the convent bell, deep and mournful, and she knew she was too late.

Véronique opened her eyes. The ten bells of Notre-Dame were ringing, creating a rippling sea of sound. The room was swathed in the pearl-grey glow of dawn and her body was sheathed in sweat. The chimes and screams of her dream gave way to the growing sounds of the city: wagons rumbled through the narrow street, riverboat men were calling, handcarts clattered, shop shutters rattled, water was slopped on stone. Paris – this place of ten thousand souls – was already awake and at work.

Véronique lay on her side and hugged her knees to her chest. She was here now. Safe. She did not have to rise, half asleep, for Prime, did not have to kneel, bone to stone, head bowed, under the disapproving stare of Soeur Cécile. For breakfast, she would eat not thin, cold gruel, but warm *petit pains* with fresh churned butter and a cup of rich hot chocolate. Yet for all that, she was uneasy, adrift. At the convent she had at least known what was expected of her. Here, things remained opaque.

It was a fortnight now since she'd returned home and her father had begun her long-awaited training. Each morning, he would teach her: anatomy, physiology, drawing, mechanics. He showed her the workings of some of his automata, teaching her and then testing her on how each part was made. They studied from beautifully illustrated books with spines of silver and gold, Véronique copying the illustrations, labelling the body parts and memorising each tendon, each bone, each vein, to recite them back to her father. The grey drawings and silver bindings had begun to bleed into her dreams: moving anatomised men, dissected women, their eyes open, their wombs empty.

Véronique picked up the worn copy of *Automata and Mechanical Toys* that she kept by her bed and slid back beneath the blanket. She thumbed the worn pages until she reached the mechanical Christs and the leering clockwork devil, black tongue protruding from his mouth. Her father had given her the book just before he took her to the Convent Saint Justine all those years ago. 'They'll be able to look after you there far better than I can here,' her father had claimed, with the groundless confidence of adults.

She remembered, as a series of flickering images, packing up her books and dolls, the drive through the night, then the sight of the convent, a medieval stone fortress, its walls stark against the milky dawn sky. When they entered, the quietness of the place had closed itself around her: no shouts, no vehicles, no ticking clocks, just the occasional clicking of rosary beads and the shuffle of feet on stone.

She remembered too the Reverend Mother Abbess as she'd greeted them, a flabby smiling face, white as fish flesh, squeezed into her wimple. 'Welcome, *ma fille*. We're glad you have joined us. I'm passing you into the safekeeping of our pupil mistress, who looks after all the boarders. This is Soeur Cécile.' And another woman had stepped forward, taller and thinner, her eyes sharp as flint, and held out a bony hand.

'Véronique, you must be a brave girl,' her father had whispered, pushing her forward, when she remained frozen by his side. And she had taken the woman's long, cold fingers and suspected bravery was only one of the qualities she'd need to survive in this place.

'We teach them humility,' she heard the Abbess telling her father as they walked from the room. 'We teach them the proper ways of this world, with kindness, compassion and love.'

But the grip on her hand had not been kind as the woman took her away from her father. The expression on her face had revealed no love as she showed Véronique up the stairs and into the long dormitory where she must sleep with the other girls. 'That bed,' she pointed, and it was the last one of about twenty, lined up against the walls as in a hospital or morgue, the girls' sleeping forms shrouded within them. Each bed had a name plaque at the end, save for hers which was blank and white. On the bed next to hers the plaque spelled out the name 'Clémentine'. Véronique placed her box at the foot of her bed and saw with a start that though the girl lay still, her brown eyes were open, shining.

Véronique had heard footsteps outside and looked down from the window to watch her father walking back to his carriage; had watched with cold understanding as the carriage rolled away.

Now, she ran her finger over the picture of Theophilos'

tree of gilded bronze, its branches filled with clockwork birds. Miles from home, from anyone she knew, from anything she understood, the book of automata had become not just a story book, but a friend, a guide. Daedalus and his breathing statues, King Alkinous and his gold and silver watchdogs – they'd been a portal to a world outside the convent, a glimpse of magic beyond the cold dormitory, the sour smell of the refectory, the boredom of constant prayer. She understood that many of the nuns had found a sort of sanctuary here – that their lives outside had been harsher, poorer, more perilous. For a child, however, this was no place at all. Certainly not a child like her, not used to the company of other children nor the idea of fettering your mind. Aside from Clémentine, her only friends were the golden birds and silver mechanical mice of her imagination. As the weeks and then months passed, Véronique resolved that she too would make such things as she saw in the book; she would build creatures that would move and fly and astonish. She would build her own clockwork universe.

Véronique's father had written to her every fortnight. His letters had been full not of his life in Paris, but of the automata of past centuries and of the machines he himself wished to build. He'd sent her drawings and explanations of whichever project he was working on: a mechanical owl, a moving hand, a troop of tiny drummers. He'd set her riddles too, mysteries she must solve, puzzles to which she must find the answers. In a place that was often lonely and sometimes frightening, his letters had been a lifeline.

Then, when Véronique was older, he'd begun to send her more complex reading materials that he must have insisted be passed to his daughter: books on clockmaking and mechanics; anatomical books that he told Madeleine she must copy from to understand the human form; charts and plans and diagrams. The other girls had giggled at the books, at the naked, hairless figures, but Véronique had known what those figures were: they were her opportunity, her way out of this stone-walled tomb. She had set herself to work, copying each illustration again and again until they resembled the original. While the other girls practised at piano or sewing, she had learnt each body part, each tendon, each vein. She had studied the essays and explanations her father sent her and she had learnt Greek and Latin better than any of the other pupils, because for her those languages were alive – the languages of science and anatomy. 'Learn as much as you can,' he told her on his parlour-room visits. 'When you're seventeen, if circumstances permit it, I will teach you myself.'

Soeur Cécile had scoffed at this idea. 'The boarders here are destined for marriage, or the veil. And yet you think you're somehow different, do you? More special than everyone else?'

Yes, she thought, because she already knew she was something the other girls were not.

'Your father will do as all the other fathers do and either marry you off, if anyone will have you, or return you to this very convent as a novice *religieuse*.'

Oh no, he will not, Véronique had said to herself, *because I will make him see that I am valuable. Because my father is not like other men.* More than that, she could not come back here, would not be forced to take the habit. It was one thing to choose this life for oneself; it was quite another for it to be used as a prison. Véronique knew what could happen to those locked in here against their will as a punishment or by necessity. She knew because she saw it every day in the face of Soeur Cécile. *Forgive us our trespasses*, Véronique thought, *because sometimes they are deserved.*

She'd hoped that her father would come to collect her from the convent himself, but instead he'd sent Edme, whom she recalled with vague fondness from her childhood, but who'd grown heavier and more severe, and who no longer knew what to say to her. They'd travelled together back through the countryside, arriving in Paris before nightfall, and Véronique had been silently unnerved by the sounds of the city: by the barking of dogs, and shouting of men, the bells from a hundred churches, the crack of whips and neighing of horses. She'd recoiled at the smell: sulphurous, human; the stink of tanneries and day-old fish, of stale beer and congealed blood. Most of all, she'd been horrified by the beggars with their gaunt faces, war-mutilated bodies and pathetic shreds of clothing, running behind carriages, crouching in doorways, holding their arms out for food. There were so many of them, and so many so young. She hadn't remembered that about Paris. She was a stranger here.

There had been no warm homecoming, exactly, no fond embrace, but then that wasn't her father's nature; nor, in fact, was it hers. *We will get used to one another*, she told herself. *The way will become clear soon.*

Footsteps in the hallway. Her maid, come to dress her and tend to her toilette in that strange blank way she had, as though the life had bled out of her long ago. Véronique closed the book of automata as the door creaked open.

'You're awake early, Miss.'

'Bad dreams.'

Madeleine said nothing to this, only set down the silver tray and laid the coffee things out. Véronique had hoped that her chambermaid might be a companion of sorts, someone to help her navigate the maze of Paris, but so far the woman was a closed box with her clever grey eyes, her copper hair almost hidden beneath her mob cap, and that odd mark that ran down the side of her face. It seemed strange that her father should have chosen someone scarred to be her *femme de chambre*, but then she knew so little of what was normal in this strange city of glitter and squalor, where even the people's accents were harder, harsher, and where no one said what they meant.

Well, she would make herself say what she meant, and what she'd been wanting to ask. 'Madeleine,' she said, as the maid poured cream from the little porcelain jug. 'What happened to your face? To cause the scar, I mean?'

The maid straightened herself slowly, keeping her eyes to the floor, and for a moment Véronique thought she might not answer her.

'I fell, Miss. When I was a girl. I fell onto the fireplace. And the tongs, you see, they were still burning hot. My mother doused me in cold water after, but already the mark was made.'

Yes, Véronique could see it now: the indentation of the metal on the skin, the slight puckering at the edges. Yet there was something about Madeleine's explanation that didn't quite ring true. She saw for a moment Clémentine's white back, the lines of red scored across it.

'I'm sorry,' she told her. 'And I'm sorry to have asked. Only, you see, I'd been wondering.'

At last, Madeleine looked up and met her stare, unblinking. Her eyes, Véronique thought, were beautiful: a deep agate grey.

'That's all right,' she said blandly. 'Many do wonder, I think. They just don't have the courage to ask. Now, will you be wanting your breakfast up here, or will you come down to the breakfast room?'

'I'll come down, thank you, Madeleine. So I should really get dressed.'

As the maid helped her out of her night shift and into her slip and gown, Véronique felt that some of the tension between them had dissipated, and she realised that the maid had been waiting for this moment – had been waiting for her to ask. And like anyone waiting to give a speech, she'd prepared her words carefully.

After breakfast, Véronique collected Franz and made her way towards her father's workroom. As she approached she heard a noise: a metallic thud. Then came another, and another, and from behind the workshop door appeared an astonishing automaton rabbit, its ears twitching. It was of course the rabbit on which her father had been working, but now, instead of the pieces of scrap metal she'd seen sewn together to form the prototype, it was like a real living thing. It hopped. Its limbs were a miraculous contortion of silver and steel. Its eyes were flashing rubies. It hopped its way over to the bottom of the staircase, wiggled its ears again, more slowly, then stopped.

For a moment Véronique too remained motionless, her heart pounding. Then she put out her hand towards the rabbit.

'Don't touch it!' Doctor Reinhart appeared from the doorway in his thin leather gloves and stooped to inspect the rabbit. His brow was creased. 'It is supposed to hop for longer than that. I will need to adjust the mechanism.'

His eyes rested on Véronique. 'You are surprised. You did not think I would do it?'

'Oh no, no, it's not that. It's simply that ...' She put Franz down next to his metal brother. 'Well, it moves just as Franz does. And yet it's something entirely different.'

He nodded. 'That, Véronique, is why you must work with the real animal. Observe it, learn its every move, capture its essence. This is why we must observe the world in detail and recreate it. You understand?'

'Yes.' But she didn't understand why, in that case, he had kept her for most of her life shut up away from the world and its wonders.

'We are not there yet, however. Further experimentation is needed. Bring Franz and you shall watch as I work further on him. We will continue your lessons this afternoon.'

For the next two hours Véronique watched, just as she had every day for the previous fortnight, as her father continued his astonishing work, filing and soldering, riveting and bolting, fitting and adjusting until each piece moved exactly as it should. The pieces with which he worked were minuscule: tiny silver screws that had been hammered through a sieve, copper cogs no bigger than a flea. Every so often her father would point something out to her or question her, to check she was paying attention. 'You see? You are watching? You understand? I am making sure it is all going in.'

Occasionally she realised he was staring at her as one might at a laboratory rat, seeing how it was performing. So far he'd expressed no frustration with her, had spoken no harsh words, but she felt his scrutiny at all times, and worried that she wasn't living up to whatever expectation he might have had of her. That, like the silver rabbit, she wasn't working quite as she should.

And though she found it hard to admit it to herself, her father was not the hero she had constructed in her own mind. Over the years she'd assembled a man made of different pieces: the surgeon who'd cut her, as a babe, from her mother; the father who'd made her a moving monkey, the teacher who'd kept her alive through the dead years of the convent with his letters and stories and sketches; the innovator, the educator, the creator. Yet in the flesh he was often inscrutable and distracted, never touching her, rarely asking how she fared, but focused on his creations. When he did engage her in conversation, it was not to ask her about her time at the convent (for which she was grateful) or to tell her about his life in Paris, but to theorise about philosophical and scientific matters: whether it was possible to reinvigorate dead matter, whether the soul was part of the body, whether one could change how blood flowed.

At that moment the doorbell rang, a rare sound in this house of few visitors. A few moments later Joseph appeared to announce the arrival of 'Claude Nicolas Lefèvre'.

'Well, bring him through!' Reinhart answered shortly.

Lefèvre. It was a name Véronique knew well. Her father had written of the man in his letters and mentioned him in their lessons, speaking of him with something approaching warmth. That was unusual. From what Véronique had seen so far, her father admired almost no one. Most of the visitors they had to the house were workmen delivering the tools and equipment her father needed, and occasionally the goldsmith, a fragile-looking man who created some of the most extraordinary things Véronique had ever seen: gilded birds and silver beetles that seemed to have been moulded from the real creatures. Doctor Reinhart never complimented the goldsmith on his work. It was either what he wanted, or it wasn't. By Lefèvre, however, he appeared impressed. 'An original thinker,' he had told her. 'A surgeon not stymied by precedent and convention. Hugely ambitious. Willing to take risks.'

She'd expected someone severe-looking, someone wizened and austere. The man who now sauntered into the hall, however, had nothing of the severe surgeon about him: he was a scruffy, pink-cheeked man with bandy legs in velvet breeches and a rounded stomach straining at the buttons of a grass-green waistcoat. He'd removed his wig and stuffed it into his waistcoat, so that the tail trailed from the pocket like a hidden mouse.

'Reinhart,' he said, his voice thick as treacle, 'you are a rogue. You have been keeping from me the fact that your lovely daughter has returned to you.' He bowed to Véronique. As he smiled at her, his small, shiny black eyes were almost lost in the creases of his face. 'Welcome home, Miss Reinhart. I knew you when you were but a babe, though of course you won't remember my haggard old visage. Reinhart has kept you shut away all these years.'

Reinhart regarded him evenly. 'You knew perfectly well Véronique was returning to me this winter.'

'Did I?' He rubbed his wigless head. 'Well, I grow forgetful these days. Too many projects. Too many clients. Too many unreasonable demands.'

'How is your pupil?'

'Inquisitive, excitable, not as quick as yours, I'll wager.' He smiled again at Véronique, then raised his bushy eyebrows at the rabbits. 'And who are these? New friends?'

'I am creating another toy for court,' Reinhart explained.

Lefèvre chuckled. 'One day, Max, you will be given an interesting commission and not just another request for an animated ornament.'

Reinhart regarded him with mild annoyance. 'I do not regard them as ornaments. They are a way of learning about life. Come – to the workshop! Véronique is going to help me improve it.'

'Is she, indeed?' He walked beside her as they followed Reinhart along the hallway, Véronique holding Franz, Reinhart holding the automaton. 'So what is your father training you in, Mademoiselle?'

'Well, everything necessary to become an automaton-maker.'

'Is he, indeed? And I'm sure he is taking you at quite a pace!'

Véronique kept her eyes to the floor. 'There's a great deal to be covered.'

'Of course, of course.' They had reached the workroom. 'Reinhart, you are teaching her ...?'

'Anatomy, physiology, clockmaking.'

'Ah well, you should let me teach her anatomy. You are an exquisite *mecanicien* and, I would wager, a lousy tutor.'

Reinhart frowned. 'No, I don't think so. Where would you

find the time, what with your research, your lessons with the King?'

'You teach the *King*, Monsieur?'

He laughed at her astonishment. 'I know I don't look much like a royal tutor, but yes, my dear, I give him certain lessons at Versailles. It's his latest whim, and who am I to refuse? It is a way to other things.'

'What do you teach him, exactly?'

'Well, he's very interested in the workings of the body, just as he is the machinations of clocks and other instruments. I have told him in fact, Reinhart, that he should come and see how you work here. I've no doubt he'd be charmed by what you create. He's much taken, at present, with the idea of animation – of how the body moves and replicates itself. I have been talking to him of electricity, how it can stimulate the movement of muscles in beasts.'

'You do not let him loose on live animals?' Reinhart asked.

'Lord, no. Dead frogs, mainly. Eels. The occasional cat. Anything from his menagerie that dies. We had a mongoose last week. And we are forever having to try to revive his beloved chickens. They sometimes fall off the roof, you see,' he told Véronique in a tone of mock confidence, 'which upsets him terribly. Or they fall prey to his many dogs.'

Véronique suppressed a smile. 'But why, Monsieur, does he keep the chickens on the roof?'

Lefèvre shook his head. 'Perhaps to keep them away from the dogs.'

'And how,' Reinhart asked, 'does he think he will revive these animals?'

'He has some ideas, but they are somewhat ... vague at present. He wishes, you see, to be a true scientist, a man of learning and innovation. He thinks this is his way to greatness and cannot always be told. In the meantime, we buy more chickens.' He winked at Véronique. 'And what was the convent like? Not much interest in anatomy there, I expect.'

For a moment, she thought of the body of Christ, depicted to anatomical perfection, the blood gushing from his right side, his forehead punctured with thorns. She smiled wanly at Lefèvre. 'I'm very glad to be here, Monsieur.'

He looked at her thoughtfully. 'And we are glad you have returned. Reinhart, call us up some mulled wine, will you? I'm chilled through.'

'Very well. And then there are matters we need to discuss.'

A look passed between the two men. There were things, evidently, that her father did not tell her, would not tell her. No matter – she would find them out. When you grow up with no adult to truly protect you, you grow up sharp, crafty, watchful.

'Véronique,' her father said, 'ask the maid to fetch us some wine and biscuits. Then start on the William Harvey texts I gave you. We will discuss them later this morning.'

Véronique gave a brief curtsey to Lefèvre. 'Monsieur.'

'A pleasure to see you again, Mademoiselle. I hope I'll convince your father to let me assist you with your studies. Indeed,' – he lowered his voice – 'if I may, I would recommend that at this stage you read not Harvey, but Lyser. His explanations are clearer, his illustrations finer. He has a better understanding of the spark of life. Reinhart, your copy of the *Bibliotheca Anatomica,* if you'd be so kind?'

Reinhart glared at him, then pointed at a shelf behind them. 'Third row down. In the middle.'

Lefèvre ran a plump finger along the spines, coming to rest on a green and silver-bound book. He took it down, flicked through it and inserted the ribbon on a certain page. He handed the book to Véronique, saying quietly:

'Begin here. He should have started with this, but he forgets that not everybody knows what he does. I'm sure you're doing much better than you think.'

After asking Madeleine to bring the refreshments, Véronique wandered up to her room to read the book that Lefèvre had given her. The morning sun glinted through the window-panes and fell onto the yellow pages of the book, showing the interior workings of the female: the womb, cut in two. She struggled to concentrate on the text around it, thinking instead of her mother, whose own womb had spelled her death. The only hint Véronique'd had of her life was the dark portrait of a sombre-faced young woman that hung on the wall in her room, and a few books with her mother's name inscribed inside the covers with a small and perfect hand: *Honorine Elisabeth Reinhart*. She'd been only nineteen, little older than Véronique herself was now. Had her life been happier than Véronique's? Had her husband treated her with greater warmth than he did his only daughter? Sometimes Véronique wondered whether it was from her mother that she'd inherited the dark rage that sometimes welled up from within – a torrent of anger at the whole world and at her own self within it. She couldn't talk to her father, though, didn't dare leap the expansive gulf that she sensed existed between them.

Abandoning any attempt at concentration, Véronique closed the book, took off her shoes, and tiptoed back down into the hallway to listen at her father's door. They seemed to be discussing some book Lefèvre had been reading.

'He argues, quite convincingly I think, that life is a vital, physical force that we can almost measure, increasing and decreasing to very precise laws.'

'Well, of course we can measure it. This is hardly new.'

'Perhaps not to you, Max, but to recognise life as a physical force similar to other forces is a huge advance for many. I feel this is the future, you know. It should be the focus of my research.'

They went on like this for some time and Véronique grew bored, her mind wandering, and was about to return to her room when she heard:

'For once, Claude, your views are decidedly behind the times. The salons of Paris are full of women discussing the sciences.'

'And the laboratories are still composed almost entirely of men. But Max, I think you misunderstand me. I'm fully in favour of the girl having an education. Indeed, I've offered to assist.'

Véronique froze only for an instant, then moved closer to the door.

'No, no,' Lefèvre's voice continued, affable but firm. 'It's not that, Max. I'm merely asking you about what you intend for her. You can't mean to keep her cooped up here with you for ever, and she surely can't follow in your footsteps, can't study at the institutions you have. She's a woman, or shortly will be. She'll be expected to marry.'

Véronique felt the coldness settle in her stomach like a stone. Why couldn't she follow in her father's footsteps? She'd known she might not study in the same places that a man would, but surely she had the best tutor in the world here in this very house. She heard only a mumble from her father and crept still further forward to listen better.

'I know what society will expect, Claude. I am not entirely divorced from the outside world.'

'Well, I do sometimes wonder. She's young now, pretty and vivacious. You should be making a match for her.'

'She is merely seventeen. I have no desire to marry her off to some crowing courtier. Better that she returned to the nunnery than that.'

Véronique's breath stopped in her throat. No, better anything than that.

'But does she not at least deserve to know what your plan for her is? She may think you intend her to be your apprentice, but that she cannot be, or at least not in any formal way. She cannot set up as an automaton-maker herself, can she, so why train her up as such?'

No answer, or none that Véronique could hear. Was Lefèvre correct? Véronique felt fear worming its way through her. What if, no matter how hard she worked, she couldn't truly be her father's apprentice, never be a *mecanicien* in her own right? What then would be her future? And why, then, had her father brought her back – why had he said he would teach her?

Further conversation she could not make out, for they seemed to have moved to the other side of the room, and then: 'Oh she has, I think, considerable potential, but only time will tell. It hinges on how she responds to certain tests and arguments that I will put to her. Now tell me what you think about the movement of this rabbit? Do you think it could be improved?'

Véronique backed away from the door, feeling her heart pulsing against her ribs. What if she failed in these tasks he set her? Would she be married off to some elderly Count, discarded like a broken toy? Or worse, far worse, would she be made to return to the convent, forced to take the veil against her will? Soeur Cécile's voice was like an asp hissing in her head: 'We will say goodbye, but not farewell. For I doubt your father will abide you for long.'

Véronique walked slowly back up to her room. The old feeling of dread had returned, as though it had never left her. From her bureau she removed the small oval box her father had given her on her twelfth birthday, crafted from three-colour-gold and pink enamel. On its lid was a figure of a girl, her arms held out as though balancing, her feet upon a golden bow.

She took a key from the side of the box and inserted it into the lid, then wound it. The tinkle of music began, like the trill of a bird, and then the girl on the box began to jump, moving up and down above her golden perch. 'It's a rope dancer,' Clémentine had said when she'd shown it to her. 'You know, like they have at fairs.' Véronique had never been to a fair, never seen a rope dancing girl, but she could imagine her perfectly because the golden girl was so real. She jumped just as a real girl would, her knees bending, her tiny feet pushing her off into the air, like magic. 'We'll go to the fair together,' Clémentine had told her. 'Just as soon as we escape.'

Clémentine was forever talking of escaping. But she had nowhere to run to; while Véronique had managed to kill only one of her parents, Clémentine had lost them both. 'There'll be a way,' she'd whisper to Véronique after the candles had been extinguished. 'I won't be here for ever.' In that, of course, she was right.

Gradually the girl slowed and ceased to dance. The music grew discordant and warped, then stopped. Véronique ran her finger over the golden girl. 'Well, I'll just have to prove myself, won't I?'

Whatever tests her father set, she must pass them. Whatever he asked of her, she must do. She could not return to the convent, for to do so would mean death.

5

Jeanne

Jeanne was imagining falling but she sat, statue-still, staring down from her window at the immaculate lawns of the Parterre du Nord where mermaids and cupids were frozen in stone and where courtiers in brocaded coats sauntered about the topiary avenues. From this distance they looked like a multitude of metallic beetles scuttling, and at that precise moment she could have crushed them all beneath the heel of her tight pointed shoes – those silly little men and women with their pretensions and their ambitions, their fine lace and false hair covering lice-bitten scalps and spotted skin, fawning and flattering and meaning not a word of anything they said. But then she was tired. Cold sweat caked the white lead paint to her face and though she'd bathed again that morning she could still smell the taint of sickness beneath the jasmine oil and above the copper tang of blood.

'Surely that's enough.'

'Just a few drops more, Marquise. We need to reduce the acidity.' Doctor Quesnay was at her side, holding the white porcelain bowl into which the blood ran red from her arm. He had bled her only a little these past few days, but she still felt hollow as an eggshell. Her strength was running out.

'There.' The doctor pressed a piece of muslin to her skin, set the bowl on the table and returned his lancet to his leather case. 'Now you must rest. For a good while.'

She raised her eyebrows at him, but not did not reply. He knew as well as she did that there was never any rest in Versailles, this gilded carousel where there were always supper parties to arrange, ballets to organise, building projects to supervise, letters to write, dignitaries to impress; where she must pay constant court to the Queen, the Dauphin, the Dauphine, the Mesdames, and a hundred other people who hated her. Where a person was judged entirely by their dress, their family, their money, their manners, and where no matter what she did, she was never allowed to forget that she was daughter to a penniless fraudster.

'I am quite serious, Marquise. If you wish to preserve yourself, preserve your youth and health, then for at least an hour each day, you must sit still, not writing, not reading, not planning, not attending to the King and his desires.'

'My youth is gone, François, you and I both know that, and rest and exercise will not cure me. In this supposed new world of science and wonder, surely there is some remedy.'

'Marquise, I wish that I knew of one. I continue my investigations. But in the meantime—'

'Yes, yes. Very well, François. Leave me now for this supposed curative rest.'

She did not stand up as he left the room but remained in her seat by the window, silent, thinking. They were writing an encyclopaedia filled with the world's knowledge, picking apart the universe and examining it piece by piece. Somewhere, there must be a solution. It was just a matter of finding it. Miette, her pet monkey, also sat quietly, chewing some candied pear. For a few minutes there was only the silvery tinkle of the fountains, the drone of the beetle-like courtiers. When she closed her eyes Jeanne was a child again, back in her room at her mother's house, listening to the chink of glasses and distant laughter, muted sounds of merriment below; back when her future seemed like an unfurling blossom, Louis merely a handsome prince, an almost mythical man; when she was just Jeanne Poisson, the prettiest girl for miles around. 'Her destiny's written clear in the cards,' the fortune teller had told them. 'Your little daughter will be the King's *maitresse en titre*. She'll have the world in the palm of her hand.'

But for how long? The woman had not said that. She had not hinted at the many forces that would attempt to pull Jeanne from her royal perch and push her into the Paris mud. She'd made no mention of the circling courtiers who every day accused her of something new, something darker; nor of how her own body would fail her before she was even thirty years old.

Jeanne's gaze was on the rooves of the Trianon, shrouded by the forests of Marly, when she heard the sound of footsteps in the corridor outside. She knew it would be him from the familiar tread of his heeled shoes, the swish of his heavy cloak, and it was though a wire tightened around her chest. She composed a smile and turned to face him. '*Mon coeur.*'

'You will never guess what had caused it.' Louis was breathing heavily from his climb up the stairs, his face was moist with sweat. He tore off his cloak and threw it down, walked to the table and poured himself some of the sweet wine that she always kept out for him.

'Caused what, dearest?'

'Tourier's tic!'

Tourier, the King's foreign minister, a man with a long skull and a strange fluttering of the eyelid. He had died two days previously, found slumped on a divan in the Palace gardens.

Louis sat heavily on the couch, wiped his face and sipped on his wine. 'It was his liver, you see. His liver was horribly diseased.'

Jeanne swallowed. 'You went to see him anatomised.'

'La Peyronie performed it himself. It turned out that he must have been suffering from some complaint of the liver for many months. The whole thing was black, like a piece of decayed meat.'

'And how do you know that was the cause of his tic?'

Louis gestured with his free hand. 'It's a complex matter. I can't explain it entirely, but La Peyronie was fairly convinced of it. Is it not incredible that our innards control how we behave, how we appear?'

'Indeéd. Remarkable.' As she spoke, Jeanne considered how the gilded skin of Versailles concealed the bubo forming beneath, gradually swelling, gathering pus; how the meticulous courtesy and rigid etiquette of the court masked the malignancy running beneath. How her own appearance, painted and perfect, belied her interior decay. She felt the silk of her dress clinging to her armpits, despite the powder. 'Do you not think though, Louis,' she said carefully, 'that you should indulge some of your other interests also? After all, you are already taking your anatomy classes, and now ... this. Horology, perhaps. Mechanics. You were once so fascinated by them.'

'I still am, Jeanne. I still am. Because all these things are connected, do you not see? The movement of the body, the movement of machines.'

'I do see.' What she saw was that this obsession with science and medicine was his attempt to be modern; his attempt to outshine the Sun King. Louis XIV might have created Versailles and the court that surrounded it, ruling in splendour for seventy-two years, inspiring fear and love and devotion, but he had had, Louis insisted, petulant as a child, a rather poor grasp of science.

'This is precisely what Lefèvre has been teaching me,' the King continued. 'He says that one day they will create a moving machine that bleeds. Indeed, your Doctor Quesnay has designed an anatomy to test the therapy of blood-letting. We were discussing it only last week.'

'Were you, indeed?' She touched her arm where she had been bled. She had made clear to Lefèvre that the focus of these classes was to be life, not death; she thought he had understood. 'Shall I have du Hausset bring some food for you? Some fresh peaches from the greenhouses, perhaps? Some lemon tartlets?'

Louis shook his head. 'No. I have only a few minutes before I must talk with Choiseul about the war on the wretched vagrants and the arguments with the even more wretched clergy. It's all so tiresome. Now, come here.' He held his hand out for her. 'I did not come here for tartlets of that sort.'

Jeanne rose from her window seat to join him on the couch, trying to remind herself of when their lovemaking had been a thing of pleasure, when she had viewed him with anticipation, not fear. She willed herself not to flinch as he lifted her skirts and felt for the soft flesh of her thighs. He buried his head in her neck, not seeming to notice the scent of illness that she was sure still clung to her. He must, however, have sensed her reluctance as he said with a hint of annoyance as he parted her legs, 'Always so cold now, my little *poisson*. It's almost as if you forget I'm King.'

After he had left and Jeanne was washing herself in the bidet, her *femme de chambre,* Madame du Hausset, came in with fresh sheets and a hot posset in a silver cup, which she set down on the table. 'To fortify you.'

'Thank you, Hausset. You are, as ever, a pearl.' False, composed from grit. But then grit was needed in this place.

Jeanne sat drinking the spiced wine as the maid stripped the sheets, her nose wrinkling at the smell of sex, which mingled with the gardenias in the vase.

'I wish to speak with La Peyronie. You will take him a note shortly.'

'Yes, Marquise. In fact, forgive me, a note came earlier for you. I left it on your writing table.'

Jeanne sprinkled rosewater on her wrists and bosom, then moved to the table in her study and took out her quills and paper. The surgeon had no place inviting Louis to attend postmortems. He knew all too well about Louis' morbidities. It was not Louis' fault, of course – first his parents, then his brother, his last mistress, the one before that – but it hardly did to encourage this fixation on death. It did nothing for his melancholic moods and it was always she who was left to manage them.

First, however, she turned to the envelope that Hausset had left on a golden tray, the paper rich in quality. An invitation, perhaps, or yet another request for favours. As she opened it, however, the breath caught in her throat. How had they got this to her chambers?

There is one thing in this complex court
That is in little doubt
The gilded fish begins to stink.
Her time is running out.

Swallowing down her fear, Jeanne rang the little golden bell for du Hausset. 'Who brought this, Hausset? How did it get into my rooms?'

'The footman brought it, just as he brings all your letters. I didn't think—'

'No, you didn't. I will also need you to get a letter to

Lieutenant Berryer at the police headquarters.' Jeanne took out another sheet of paper. 'And Hausset?'

'Yes, Madame?'

'Pray do not read it this time.'

6

Madeleine

The ninth of February. So said the case clock in the hallway that Madeleine now dusted, the number nine appearing below a silver crescent moon and stars. Six days had passed since she'd come to the house and still Madeleine had found no evidence of these damned 'unnatural experiments' Camille had rattled on about. She began to wonder if he'd misled her on that, and on a whole lot more besides. Though she'd inspected the workshop and the other rooms several times, had read Reinhart's correspondence and listened at his door, had asked subtle question of the other servants, she'd so far found nothing about experiments and nothing to prove what kind of man Reinhart was. Her gut, however, told her something was wrong, with the man and with the house itself. There was a sickness here, a shrouded strangeness, lurking beneath the silver and silks.

She moved on to the next clock with her cloth. 'The dust mustn't get inside,' Agathe had told her, 'or it will stop the clock working, so it's vital you keep them clean.'

She rubbed her eyes, gritty from dust and exhaustion. For it wasn't only rabbit-procuring they had her doing. It was all sorts of odd tasks that they'd no mind, Madeleine thought, to be asking a maid to carry out, on top of her already full day. She hadn't even had time to run home to Émile to see how the boy was faring. She rose every morning at half past five to fetch fuel, sweep the hearths and blacklead the range, and then, once she'd scrubbed the dirt from her chapped hands, help Edme with the breakfast and assist Véronique to dress. The rest of Madeleine's mornings were taken up with cleaning, washing and fetching whatever Doctor Reinhart had set his mind on that day. He was a strange man, no doubt about it. Brilliant, perhaps, but queer. Most days he seemed barely to notice her existence, but there were times when she caught him looking at her, or rather looking at a part of her – her face, her shoulders, her waist, her hands – as though measuring her up for a new suit of clothes, or perhaps for the hangman's noose. It wasn't just her he looked at funny either – it was his own daughter. He watched her like she was one of his clocks, to check it was working properly.

Had things been different and she a different sort of girl, Madeleine supposed she might've felt sorry for Véronique, friendless as she was; might've worried about what would become of her if the police decided Reinhart should be removed. But every time she helped her mistress – tying the girl's sash, fastening her buttons, curling her hair as she'd once curled Suzette's – she thought of her sister and the things she'd stomached in place of Véronique's studies and reading and walks to the park. Seeing Véronique lying in her comfortable bed in the mornings, Madeleine even recalled things she herself had endured – thoughts that she snuffed out like a candle. And each day, as her time ticked down, her resolve strengthened further: she'd escape Maman's house for good and create a new life for herself and Émile, away from the Rue Thévenot. *D'affranchir*, that's what the card sharps called it: to save one card at the cost of another. It was a shame that this was the way it had to be done, but life, as any girl knew, was very far from fair.

She started, the hairs rising on her arms. Turning, she saw Doctor Reinhart in the hallway, watching her closely. He stood very still and silent. For how long had he been there?

'You must be very careful,' he said slowly. 'You must be gentle with them. You understand? Treat them as living creatures. If you set them off balance they are liable to stop, and then where would we be?'

For a moment Madeleine could only stare at him, at his unblinking ebony eyes. 'Of course, Monsieur. I'll be more careful in future.'

Reinhart nodded. 'Yes. Good girl. It's important I should be able to trust you.'

Tick, whisper, tick went the house as she lay in bed that evening, watching the candle smoke curl against the wall. Despite Madeleine's tiredness, sleep didn't come easily in this house of incessant ticking and hidden workings, its rich but gloomy rooms. Even when she did sleep, her dreams were full of clocks and gears and numbers, the relentless march of time. Just as she was finally sinking into sleep she became aware of a sound – something below her window. She opened her eyes. Listening closer, she heard the creak of the harness and the snort of the horse. Then came men's voices, low, but clear enough from her room, which looked directly down onto the street. She moved to the window, drew back the curtain, and peered down to see a carriage the colour of oxblood, the horse before it shifting its feet, its breath steaming in the cold.

Two men were removing a large black box from the carriage and a third man stepped forward into the lantern light to direct them. Madeleine couldn't see his face, only the top of his head, but from the awkward way he moved she took it for Doctor Reinhart. The box, she reckoned, was the length of a coffin. Here, at last, was something.

Madeleine moved over to the door, shivering in her shift. Should she risk running downstairs, or was that madness? She could hear Edme snoring steadily in her bed in the room opposite. If the cook or Doctor Reinhart found her creeping about the house then the whole rig would be scuppered. Then again, if she returned to the police empty-handed, she'd have no money, no way out, and the prospect of this seemed even worse.

There was more talking, but she couldn't make out the words, only the low note of their voices, one rising in a question, another answering. Madeleine gently opened her bedroom door and ran barefoot along the dark hallway and down the back stairs, the gleam of polished wood showing her the position of the banisters and hallway. She could hear voices in the workroom.

'It's not what I asked for,' she heard Reinhart say. 'I was most specific.' Mumbling, words she couldn't make out. Then, 'Too old.'

'You are very particular, Monsieur. We're just working with what we can get.'

Madeleine's stomach tightened. They were talking about bodies, weren't they? And not those of waxen women in cases. These men must be corpse-sellers. She knew their type, of course, living where she had; men who collected the unclaimed bodies from the hospitals and prisons and sold them to anatomists and doctors. Men who, when the legal sources ran dry, took to digging up the dead. Given these two were carrying in their dubious boxes under the cover of darkness, their wares didn't come from the hospitals or prisons, she'd wager. What they brought must be a crime. What did Reinhart mean by 'most specific' though? She didn't like that. Possibly it was to do with his hideous experiments. Maybe she was onto the truth.

'It is important for my work,' Reinhart said. 'You understand? Please do not return until you have exactly what I asked for.'

Some muttering, then the sound of footsteps, growing more distinct. Madeleine started back from the door and scurried back to the staircase, where she waited in the darkness, her back pressed against the wall, her heartbeat thudding in her neck.

The door opened and a rectangle of light fell across the hallway, then a man's long shadow across it. The two men passed out of the workshop and into the hallway, one short and thickset, the other taller, stooped, his shadow a thin curve on the wall. Reinhart came behind them, a lamp in his hand.

'Leave as quietly as you can.'

Madeleine fled quickly back upstairs. She lifted the curtain again and saw the driver flick the reins to urge the horse onward, then the carriage moving slowly forward, the horse's hooves becoming a hollow patter on stone. She lit a candle, took her writing things from her case and drew her chair up to her little table. She ground the sheet of paper until it was smooth, then dipped the nib of the quill into the ink.

'Monsieur,' she began, 'I write to inform you of all I have been able to discover in my first week at the clockmaker's house.'

She wrote quickly, telling him of the conversations she'd overheard, the letters she'd read, the visit from the men with a box. She paused then, her hand hovering above the paper, wondering what else to say, for she didn't know why they'd brought the body at night, nor why what they'd brought was wrong. '*Too old*,' he'd said. Too decayed, they must've meant. She shrivelled her nose at the thought. She'd have to keep her eyes and ears open, and hope that this was enough for now.

When Madeleine had finished her letter, she blew on the ink, dried it with sand, folded the paper, then dripped a pool of candle wax onto the letter. Into the wax she pressed the seal he'd given her, leaving the imprint of a fox.

'As you know, we normally call them flies,' Camille had said. '*Mouches*. They keep close to the wall, beady-eyed, listening. But you with your auburn hair, your light eyes, your stealth – I think I'll call you my little fox.'

*

The following day Reinhart announced that his rabbit was nearly complete. 'It's to be delivered tomorrow to the Maréchale de Mirepoix,' Edme told her as she kneaded a lump of pastry, 'ready for the Versailles ball.'

'A rabbit, for a ball,' Madeleine said absently. 'That's a new one on me.' She was thinking still of the men the previous night; the carriage gleaming darkly in the lantern-light.

'It'll be part of some costume,' Edme said, rolling up her sleeves and sprinkling flour on the table. 'They do all sorts of strange things at those balls, you know; dress in shameful costumes poor seamstresses have ruined their eyes over in the weeks before. Gods, beasts, all sorts. And then they glut themselves on foods you'd never dream of: piles of meringues that stretch to the sky, ices in the shape of swans.'

Edme shook her head in disapproval, but Madeleine could tell the woman was imagining, just as she was, what it would be like to go to the court, to dance in a jewelled costume in the landscaped gardens and to gorge on pastries and wine.

'Have you ever been to the palace, Edme?'

"Course not. I've no time for that nonsense.'

'My mother went,' Madeleine said. 'And my sister. Last year.' They'd left the house trussed up in their best like a pair of overstuffed pincushions, then crowded in with the rest of the open-mouthed public to watch the King eat his ceremonial boiled eggs.

'You've never seen the like,' Coraline had told her when they returned, dishevelled and reeking of brandy. 'Courtesans with their faces painted perfect as dolls, their skirts as wide as the door. You'd not know the difference between them and a lady, as they all dress and dine the same. They walk like they're floating above the ground, which 'course in a way they are.'

Maman hadn't taken Madeleine, of course. She'd been left behind to scrub the floors and keep an eye on her mother's property. If you wanted to look your best at Versailles, you didn't take the daughter with only half a face.

Edme, taking out a rolling pin, glanced at her. 'Well, you were wise not to go with them. They say the place stinks worse than the Seine in summer and they're all riddled with the pox.'

When Madeleine went upstairs to clean the silver, she saw Véronique standing staring at the rabbit hutch. It was empty, the crude door open, and Madeleine's heart gave a lurch. 'Miss Véronique?'

'He said he needed the skin.'

Madeleine's breath caught in her throat. 'Your father?'

'Yes.' Véronique stared still at the empty cage. 'To make the automaton more lifelike.'

Madeleine came to stand beside her, searching for something to say. She'd known, really, that they wouldn't keep the rabbit. After all, as Joseph had told her, Reinhart hadn't kept the goose that became the golden goose, nor any of the other animals. Reinhart seemed only to keep the creature alive for long as he needed it as a model (or a 'template', as he'd called it), then moved on to his next invention. Still, to skin the animal his own daughter had been treating as a pet seemed to touch on cruel. Had he not noticed his daughter taking the rabbit for walks? Not noticed the way she cuddled it? Or had he seen all that and decided that the girl should be taught some kind of lesson?

'It was only that I'd grown rather fond of him,' Véronique said.

'Yes, Miss.' Looking at her moth-pale face, she was reminded for an instant of Suzette and felt an unwanted pang of pity for the girl. 'Perhaps I could get you another. I could go to back to the same shop today.'

'That's kind of you, Madeleine, but it's only a silly thing. What need have I for a rabbit?'

Much need, Madeleine thought. *For you have no one else to love.* And then she cursed herself for such softness, for it would get her nowhere fast. If life had taught her anything, it was to keep herself closed off, hardened. Véronique might just have to learn the same.

Madeleine was still thinking about Reinhart that evening when she laid the knives on the servant's table for supper. She pictured his shining scalpel, his quick hands, the peeling back of skin.

'That's at least the second one gone in a month,' she heard Edme say. 'First the baker's boy on the Place Dauphine, now a chandler's assistant from the Rue de Calendre – a brown-skinned boy, only thirteen years old.'

'Gone where?' Madeleine asked, looking up.

'Well, no one knows for sure, that's the rub of it. The chandler himself reckons the lad's run away, but there's some saying it's the navy, taking young ones to the islands. It's happened before.'

Madeleine knew what Edme was talking about: youths sent off to the colonies some thirty years since. 'But they were prisoners weren't they, back then? Prisoners who were sent off to work on the islands?'

'Some of them, some not. All piled up together and shipped off like they were so many bundles of kindling. That's how much they value those that can't pay their way. Now here, help yourself to this.' Edme placed a pie dish on the table, the delicious smell of meat and herbs escaping from the crust.

Madeleine thought of Émile, who would right now be playing out on the street. She thought of her trip to buy him medicine. 'I heard two women in the apothecary's talking about a girl who'd not come home.'

'When was that, then?'

'Back in January,' Madeleine said. 'They reckoned she'd gone with a travelling fair.' She cut herself a slice of pie.

'And who knows if there aren't street children that've gone missing too,' Edme said. 'It's not as if anyone would miss them.'

'No. True enough.' She thought all at once of the girl in the Rue Thévenot, the shock of the empty doorway.

It was only when Madeleine tasted the pie that she realised what meat it was, and then, abruptly, she set down her fork, mumbling that her stomach was crook. Keeping her eyes averted, she spooned up some soup instead. Ridiculous, that she, who'd not cried in a year, should mind eating a paltry rabbit.

After a moment she noticed Joseph looking at her over his tumbler of wine and feared he'd guessed at her thoughts and believed her a little fool. There was no place for sentiment in the life of a servant; there was, truth be told, room for very little of anything. But then she saw that he'd left his own slice of pie on the plate, untouched. Their eyes met.

When Edme took a seat the table, Madeleine asked, 'What

about the first boy? The baker's boy? Is he supposed to have run off too?'

'That's what the baker himself said when it happened, but of course he didn't know then that others had gone. He might have looked a little harder.'

Madeleine felt a coldness in the pit of her empty stomach. 'If you don't mind, I'll go to my mother's shortly. Make sure they're keeping a good eye on my nephew.'

Edme nodded curtly. 'You do that. But don't be long, for I'll need you to do the washing.'

Madeleine found him on the steps in front of her mother's house. He looked thinner than when she'd last seen him – more pinched, greyish about the eyes – and he didn't jump up to meet her.

'Oh Émile, you've been ill again, haven't you?'

He shrugged. 'A cold is all.'

A cold, with his chest. It was always bad, leaving him breathless, tired and drained.

She sat on the step beside him. 'Maman should have told me. She should've got word to me at the clockmaker's house.' But then Madeleine should have come anyway, sooner. She felt the familiar tug of guilt. From her pocket she removed her handkerchief, unwrapping it to reveal the piece of apple tart that she'd pilfered from the kitchen: Edme's pastry, glazed with marmalade and releasing the scent of cinnamon. 'Here.'

Even this produced only a weak smile. 'Fancy. Is it all like that there, in the clockmaker's house? No wonder you haven't been to see me.'

'Now, Émile, that's not it at all. It's only that they keep me very busy. There's jobs for me to do every minute of the day. It's not easy to slip away.'

He nodded, but his grey eyes reproached her and the twinge

of guilt deepened to an ache. 'I'll come sooner again next time,' she said. 'I'll come back every few days, I promise. All right?'

No answer, but Émile took the cake from her and began, slowly, to eat.

'How have things been here?'

'The same,' he said between mouthfuls. 'Another girl came, Claudine her name is. She stays in Odile's old room.'

'That so?' She had to get him away from this place with its revolving door of females, healthy in, destroyed out; with its reeking steps and the stinking men who climbed them. With five hundred livres they could set up a pet shop of her own. Nothing fancy, but a more secure life for herself and Émile, where she set the prices, she made the rules, and where her body was not for sale.

'Émile, I need to ask you to do something.'

'What?'

She hesitated, unsure how to word it. 'Two boys have gone these past weeks and they're not sure where.'

He frowned at her, confused, and she bumbled on, too fast. 'Of course it's probably nothing and I don't want you to worry, but then again I don't want you to be careless. I don't want you to go too far from the house. Not on your own. Make sure you're with other children, or with one of the girls. And if anyone should ask you to go somewhere with them, you must say no. You must go back to Maman or Coraline and tell them. Yes?'

'Why would anyone ask me to go with them?'

'Well, they probably won't. But you mustn't, you see? Mustn't go.' She paused. 'You understand?'

Émile shrugged. 'If you say so. But why can't I come stay with you?'

Madeleine sighed and put her arm around his waist. 'I wish you could, my love, but the clockmaker wouldn't allow it.'

'But I'm a useful little machine, you said so!'

'Yes. And Maman needs you here.'

He wriggled free of her, folded his arms and shifted further along the step.

'It won't be for very long, Émile. Only three more weeks. And then—' She stopped. She couldn't tell him what she hoped for. If Maman got wind that she planned to escape, all of her plans would be scuppered. She'd seen what happened to girls who dared to cross her mother, who ran from the brothel with their false-fine clothing still on their overworked backs. Maman's agreement with the police worked both ways, which was why she was so keen to please them. If a girl ran off, she rarely got far before the arm of the law dragged her back. Madeleine would have to find a way of striking a bargain, or of disappearing altogether.

'And then *what*, Madou?' Émile's little face was turned up to hers in question.

'And then I hope things will be better for us, *mon petit*. A fresh start.'

She couldn't return to Maman's, she knew that now. Away from the house, from the girls, from the routine – such as it was – she recognised it for the half-life it was. Returning to the so-called 'Academie' would kill her completely, in one way or another. Never would she go back.

Winding her way through an alleyway on her return to the clockmaker's, Madeleine had again that sense of someone watching her – of something crawling up her spine. Yet when she looked back there was only ever a dog, its ribs showing through its mangy fur, or an old woman with her skirts hiked up, pissing in the streaming gutter. It was all this talk of a child thief, that's what it was, it'd set her mind askew. She was imagining monsters beneath the bed, as she'd done when she was a child. 'There are no monsters,' her father had promised her. But he had been lying. Then he himself had disappeared, leaving her to face them alone.

Just as Madeleine reached the steps to the clockmaker's house, a small boy darted towards her – thin faced, grubby – and thrust a piece of paper into her hand.

Madeleine saw that the writing was a rough scrawl, as though written in anger or haste. 'This is not enough. Find out where they're bringing them from. Find out what exactly he's looking for. Look closer, listen harder. Write to me at once.'

Reading that, Madeleine felt the cold breath of fear. '*Them*'. Camille meant the bodies. He meant do a better job or your escape route will be barred.

7

Véronique

'You should start, I think, with a doll.'

They were standing by her father's worktable, books and brass tools laid out on the surface together with a velvet box. He still thinks of me as a child, Véronique thought, a child to play with dolls. But then why had he taken her perfect white rabbit and stripped it of its skin? Reinhart opened the box to reveal the porcelain heads inside and Véronique saw that their faces were still blank ovals, waiting to be painted in. She lifted one in the palm of her hand and felt that it was hollow.

'What do you want me to do with it, Father?'

'I want you to make it something of your own. I want you to bring it to life. You understand?'

'I think so, yes, Father.' But what do you really want with *me*, she wondered. Why are you teaching me all of this if Claude Lefèvre was right?

'You will help me?' she asked.

'Of course. But first you must create a plan, using the prototypes I have shown you. Joseph and I must go to the cabinet-maker's. I will inspect your progress on my return.'

After he'd left, Véronique sat for some time thinking,

searching through her father's previous sketches and plans of moving dolls, looking through the pages of a book entitled, *Doll-Making: A Guide*. Then, on a sheet of foolscap paper, Véronique began to sketch a face, filling in the eyes. It should have been a joy, creating her own automaton under the guidance of the greatest automata-maker in France. She suspected, however, that this was part of the test – a test she must not fail. In fact, since the rabbit incident she'd wondered if everything he did was part of that trial, everything done for a reason. Soeur Cécile's words came into her head, unbidden: '*Remember, He wants us to be tested, Madeleine. It is that way we show Him our strength.*' She pictured the Christ in the convent chapel, his eyes running with blood.

The jangle of the doorbell. Would Madeleine answer it? Edme? Véronique hurried to the hallway, where Madeleine was already opening the front door. Into the house stepped a tall woman dressed in silver and blue, which reflected the light so that she seemed to shimmer. Véronique's immediate thought was that she'd never seen someone so beautiful. But as she watched the woman walk smoothly into the hallway, she realised that in fact beautiful was the wrong word. The woman was very pretty, yes, but it was the combined effect of her face (pale, wide-eyed, expertly painted), her long neck, her graceful figure and the way she held herself, and the clothes – oh, the clothes. Grey-blue silk and gossamer shot through with silver, a white fur cape like that of a fairy-tale queen. She looked as exquisite and as fragile as a porcelain doll, but very much alive.

For an instant, she could only stare at this woman, wondering who she could be and why on earth she would be visiting her father's house. She felt acutely conscious of her own plain dress, uncurled hair and unpowdered face. She dropped into a belated curtsey. 'Good morning, Madame. How may we assist?'

The woman was studying the clocks that ran along the hall-way, running her manicured fingers over the gilded casing of the pedestal clock. Behind her stood a short, brown-skinned girl, presumably her maid. The woman turned and faced them, giving them a radiant and practised smile. 'I've heard that Doctor Reinhart makes clever little trinkets. Self-moving dolls. Is that right?' Her voice was as smooth as cream.

'Yes, Madame,' Véronique said, collecting herself. 'Allow me to show you some of the things he has already created.' She led the woman through to the room that served as a shop, opened the cupboard and removed from it the now finished silver spider. 'It moves if you wind it.'

The woman took it in her hand and laughed, apparently delighted. 'How marvellous! He made this?'

Véronique nodded and then couldn't help adding. 'I helped him.'

The woman raised her eyebrows and seemed to reassess her, running her eyes over her face and figure. 'You're his daughter.' It wasn't a question. 'I'm looking for something custom-made,' the woman went on. 'Something very special.'

'Of course. What in particular would you like?'

'I'm not sure exactly. It's a present. A present for someone very dear to me. Someone who likes to be entertained.'

'Then ... a mechanical game of some kind, perhaps?'

'No, no,' she said with a wave of the hand. 'We have games aplenty.'

'An ornament? Something to wear? If you would like to look at the brochure?' She placed the illustrated book before her.

'No. It needs to be something unique. There. On that shelf. Above the mouse. What's that?'

Véronique walked stiffly back to the cupboard and picked up an enamelled box topped with a jewelled flower. She inserted the key to wind the mechanism. Music began to play and the petals of the flower opened, revealing a dancer, arms raised, dressed in Venetian lace. The woman clapped her hands. 'How very clever. And yet—' She sighed and idly turned the pages of the brochure, stopping when she reached a picture of a mantel clock topped with a silver bird. 'A bird. He likes little birds. Yes! A bird in a box, that's what it will be. A singing bird in a precious box. Wouldn't that be perfect, Amaranthe?' she said to the maid. The girl merely nodded, her dark eyes dull.

'It must be jewelled,' the woman said to Véronique. 'Sapphires. Diamonds. Opals. Emeralds. It must be the most beautiful bird your father has ever made. Do you understand?'

'Yes, Madame,' Véronique said curtly. She didn't like being spoken to as though she were a servant, as if she were somehow lower than this woman. She wondered how Madeleine bore it all the time, being treated as some form of sub-species.

'Good.'

'In order to purchase the materials—'

'Yes, yes.' The woman removed a velvet purse from her dress and laid out a pile of golden louis and notes. 'Amaranthe will return in a week to see how you are getting on. You'll let her know if you need more.'

Véronique remained silent, staring at the pile of golden coins. There must be a thousand louis there, maybe more. What kind of client would pay unlimited amounts to someone they've never met before? Looking closely, she saw that the woman's eyes seemed too bright, feverish, and that the veins stood out from her snow-white neck.

'As I said, it's needed as a gift,' the woman said. 'For a fort-night's time.' There was a note of challenge in her voice.

Véronique met her gaze. Was it possible to make an auto-maton box in a fortnight? She had no idea, but she wouldn't tell her that. 'Well, then we will have it ready for you. Who may I tell my father has requested this gift?'

'You may tell him that my name is Madame de Marinière. And you may tell him that if his gift pleases me, I will be commissioning more.'

As the woman returned to the hallway, she looked at Madeleine and gave a curious smile. Then the doorbell tinkled again and she was gone, leaving only a faint scent of jasmine in the air. Véronique looked at Madeleine and for a moment they stood, eyes locked. Then, unaccountably, they both began to laugh.

★

'Two weeks? You agreed to have it ready in two weeks? What in the name of God were you thinking?'

Véronique had never seen him angry before, not truly. His face was leaden, his eyes dark fire. She dug her nails into her palms. 'The woman gave us gold, father. Hundreds of livres. And said we could ask for more. She's clearly someone of real importance.'

'None of this enables me to create an original automaton box in a fortnight.'

'I'm sure you can. I know you can. I will help. We can order the parts in from the goldsmith, from others.' She heard the fear in her own voice.

Doctor Reinhart was pacing the workshop, pulling things out of drawers. 'And what am I to tell my other customers? I have orders to complete!'

'I don't know, Father, but I didn't feel I could refuse. She's someone, this woman. She could be your best customer if you impress her.'

'How am I to do that in a fortnight? I am a craftsman, I do not simply churn out goods. *Pardieu,* have you learnt nothing these past weeks?' He went to his bureau, took a sheet of paper and began to write. For an agonising moment Véronique thought he must be writing to the convent, to tell them to prepare for her return. She had ruined everything, simply because of the way the woman spoke to her. But then he said: 'We will need axles, gears, levers and cylinders of the smallest size. I will need to work with a box I have already started and adapt it. I will need the enameller, the gem-dealer and the spring-maker at once. Joseph, take this. I need you to go first to Monsieur Villiers. Time is of the essence.'

Joseph stepped forward. 'Yes, Monsieur.'

'And you, Madeleine. You will take this note to the gold-smith's. You will impress upon him that this is to take priority over all other work.' He took out another piece of paper, dipped his quill in ink and wrote a few words, then used sand to dry it off. 'We will need a bird, too. At once. A bird in a cage. The most enchanting and magical bird you can find. We must work with the real thing.'

'Yes, Doctor Reinhart,' Madeleine muttered, stepping quickly away from him.

'I'll go with her,' Véronique said, wanting to appease him, and wanting more than anything, to leave. 'We'll find something exquisite.'

'Very well. Go. But then I will need you here to help me.' He looked at her, but it seemed more that he looked through her. 'Since you have agreed it, we will make a bird for this Madame de Marinière, and it will be the finest toy she and her mysterious friend have ever seen. We will give them a little surprise with it, yes?' He chuckled: a low throaty sound.

Véronique did not like it.

With barely a word to each other, the girls put on their cloaks and hats and set off for the Quai de la Mégisserie, walking to the tip of the Place Dauphine and through to the Pont Neuf, the pulsing artery where the brightest and darkest veins of Paris flowed together as one: song pedlars, fish vendors, quack doctors, snake-oil dealers, all singing and shouting above the clatter of wheels and hooves. The smells mingled into a brew to make you choke: horse manure, animal skins, human sweat, lye. Across the street, the great bronze monument of Henri IV astride his horse provided a shelter for numerous beggars, some lacking limbs or eyes, some covered in hideous sores, some holding out cups, some bent in prayer.

Since her return to the city, Véronique had ventured out only a few times, mostly in the carriage, and the quantity of people she now saw walking, running, swerving among the sedan chairs, fiacres, carts and horses, made her chest tight with anxiety. Madeleine too seemed on edge, she thought. Perhaps she'd seen the fury in her father and been frightened. Then again, her maid often seemed nervy, watchful, as though waiting to be caught out. Together, they made their way past the numerous tables, trunks and stalls from which people sold chestnuts and apples, tin pots and ladles, soaps and perfumes, instruments and carpets, second-hand books and third-hand clothes. The way grew more crowded, the coachmen more ill-tempered, and as they approached the other side of the Pont Neuf she saw that people were abandoning their carriages entirely and joining the melee of people on foot: women in bright satin dresses and feathered hats, men in purple jackets and high-heeled shoes, picking their way through the black mud and scattered straw, a herd of cattle driven between them.

Reaching the end of the bridge they saw that a fiacre had been overturned, its contents spilled into the road, its frame broken. Two red-faced men were gesticulating at each other, shouting, while at the side of the road a woman sat holding a blood-soaked handkerchief to her face, a small crowd looking on.

'Happens all the time,' Madeleine said, glancing at the scene, then at Véronique's expression. 'There's even a tariff for the injuries: so many livres for a broken leg; so many for a broken back.' Madeleine seemed unfazed by this casual pricing of body parts. 'Folk should mind where they're going.'

They walked to the end of the bridge and onto the Quai de la Mégisserie, the point where the east and west axes met and where, as it seemed to Véronique, all humanity and its beasts had clotted together and now pushed past them, too close, a mass of yellow teeth, pouched and pockmarked faces, noses swollen by brandy or rotted away by syphilis. The pungent odour of unwashed bodies, tobacco, stale wig powder and unfed stomachs pressed in on her, blending with the stench of the Seine, so she could barely breathe.

'Not far now,' Madeleine assured her, taking hold of her arm to steer her away from the traffic and piles of refuse. Véronique was comforted by her confidence and closeness. She wasn't frightened exactly, but rather appalled, entranced and somewhat ashamed that she'd been kept so closeted until now.

A gutter ran through the centre of the road, flowing with a foetid stew of animal and vegetable matter, and as they drew closer to the tanneries further up the Seine the smell of sulphur brought bile to her mouth. Then, above the voices of workmen and the sound of water, they heard the trill of birds. They had reached the *oiseleurs* – the bird sellers. Outside the shops, cages squirmed with canaries, pigeons, squirrels and miserable monkeys. Véronique seemed to know the area. She selected the grandest-looking of the row of shops, a sign with gold lettering pronouncing it to be the residence of 'Ange-Auguste Chateau, *Oiseleur du Roi*'. Inside, boxed into elaborate cages, goldfinches, canaries and cockatoos chattered, macaws screamed and sang. When the *oiseleur* saw Véronique and Madeleine, he took down a gilt cage and held it up to them. 'A Peruvian nightingale,' he said. 'The sweetest song you ever heard.'

Véronique looked into the cage and glimpsed a forlorn-looking linnet.

'Thank you, Monsieur, but we wanted something prettier. Something with colourful feathers.'

Within a moment he'd produced one of the macaws, bright yellow and blue.

'Very good, but we were hoping for something more elegant. Smaller.' Something to match the grace of the woman, something that they could reproduce in silver and sapphires.

The man showed them various birds – some impossible colours that must have been painted on, others with clouded eyes, seeming close to death. It was Madeleine who spotted the blue bird, sitting silently in a silver cage towards the back of the shop. 'What about this one, then?'

The man turned. 'That, Mademoiselle, is a Louisiana Sapphire. Extremely rare.' He lifted down the cage and Véronique saw that its feathers were a vivid lapis mixed with a lighter blue, its tail feathers and head lined in black and white.

'It's a jay, isn't it?' Madeleine said. 'It looks like a jay, only blue, not brown.'

The man frowned, annoyed. 'It is, but the rarest kind.'

The bird stared at Véronique with a gleaming black eye.

'How much?'

The man snatched a price from the air, a high one. Madeleine snorted and shook her head at Véronique. 'It's not worth half that,' she said to her quietly.

'We'll give you ten louis for it, Monsieur.'

'Twelve.'

'I can go no higher.' Véronique made to leave, but she knew she'd pay. She'd never seen a bird quite like it.

'Eleven and he's yours.'

She nodded, then stared again at the jay. He was, she decided, perfect.

As they walked back towards the Pont Neuf with the bird in its cage, they noticed a clutch of people gathered around a pamphlet seller. 'Another child gone!' the man was shouting. 'Read what's written here!'

Véronique approached the group, close enough that she could read the pamphlet the man held up in the air: 'Seamstress's Daughter Stolen!' the headline shouted, and then beneath it, 'Does a child thief stalk our streets?'

Véronique glanced at Madeleine, who was also looking at the pamphlet. She'd assumed her maid couldn't read, but perhaps she'd been wrong. 'You know what it says?'

The maid flushed. 'Not really, Miss.'

'It's talking about a girl being stolen. Had you heard about this?'

Madeleine's light eyes darted at her. 'People are saying, Miss, that some children have disappeared from the Île de la Cité these past weeks.'

'How many?'

'Two tradesman's boys, that I know of: a baker's boy of twelve and a chandler's apprentice aged thirteen. Maybe a girl. And there may be others, of course.'

'May be? How can it not be known?'

'Well, street children disappear often enough during the cold months.'

'Where do they go?'

Madeleine regarded her with what might have been a hint of scorn. 'They die, Miss. In the cold.'

Véronique swallowed. 'There's nowhere for them to shelter?'

'Not for long, no. They're hounded out of Paris.'

Véronique imagined a line of children, beaten back into the forests. Was that really how things worked in this city? Perhaps she should not be surprised. 'But the tradesman's boys,

Madeleine. And this girl. What do people think has happened to them?'

'I don't think anyone knows for sure. Edme says they might've been taken to the colonies by the navy, same as happened thirty years or so ago. There's some saying it's the migrant workers from the provinces, but then they get blamed for everything.'

'You think it's the navy, then?'

'Oh, I couldn't say.' She frowned. 'It could be someone else altogether.'

All at once Madeleine's expression changed, tightened. Véronique followed the direction of her gaze and saw a bacon-faced woman hurrying towards them, the plume of her hat quivering as she moved.

'Well, if it isn't my own daughter!' The woman was breathing heavily, smiling, her formidable chest rising and falling. She gripped Madeleine's arm.

'Maman. Fancy you being here.'

'A stroke of providence, certainly. And this is …?'

'This is my mistress, Demoiselle Véronique.'

Véronique nodded at the woman, who attempted a curtsey. She looked nothing like her daughter, save for perhaps for the clever, watchful eyes.

'An honour, an honour. I do hope young Madeleine here is serving you well?'

'Very well, thank you.'

'Oh, she's had a good training, has our Madou. She's a good girl.' The woman reached up and touched Madeleine's cheek, the undamaged side. 'I'm sure she's doing everything that's been asked, and that she's mindful of her obligations, eh?' Then, nodding at the cage: 'A pet, is it?'

Madeleine hesitated. 'A bird. For Master Reinhart.'

'Indeed. He likes a pretty bird, then?' Their eyes locked.

'He uses animals for his inventions.'

'Does he, now?' A pause. 'Well,' the woman said, 'I must leave you to get on. Miss Véronique, forgive me the liberty. Madeleine, do call on your old mother soon, eh? It's already been a fortnight. Your sister and I have been missing you so, and her gentleman was asking how you fare – wondering if there was any news.' The woman raised a gloved hand, then lifted her skirts and moved away, the feathers on her hat bobbing as though in farewell.

For a moment, Madeleine stood watching her retreating form, then said: 'We should hurry home, Miss. It looks like it's going to rain.'

8

Madeleine

'Full of life,' Doctor Reinhart declared as he turned the cage to survey the bird's bright feathers.

The jay, seeming to understand him, opened its beak and emitted a strange, rasping cry.

'I will have to capture that call. Yes. I must get to work on it immediately. Véronique, you will help me. Madeleine, tell Edme we will take dinner in here today.'

'Very good, sir.' Sweat had pooled at the bottom of Madeleine's spine. She felt lightheaded, as though she'd a fever coming on. *It's already been a fortnight.* Oh, and didn't she know it? A fortnight in which she'd found nothing of real substance, answered neither of Camille's key questions. That was what Maman meant, appearing, red-faced and vulgar like that – the worst actress that ever walked the boards. She meant, as she always did, do your job or you'll be for it. As if Madeleine needed reminding! She was sorely aware of how much time had passed, and how little time she had to succeed. She thought of it every time she dusted the many clocks, every time she tried to sleep.

In the past week, Madeleine had been several times through

Doctor Reinhart's workshop to see if she could find some clue, some hidden compartment or secret alcove, some hidden door concealing terrible truths. She'd turned his bedroom upside down, then righted it; she'd even searched the stables, but nix. The only new items were parts and equipment that must be needed for the making of the automaton box: cogs and gears, tiny chisels, a glass jar with a metal spike through its stopper. There'd been no further night-time deliveries, no whispered conversations, no mention of bodies or experiments. It was not, she sensed, that there was nothing to find, only that she didn't know where to look. For there was something wrong with Reinhart, something on which she couldn't quite put her finger. He didn't look at her as other men did, but like he was planning an incision; as though he'd divide her up on his anatomist's table to see how she worked inside. And he looked at his daughter in much the same way, as though he saw right through to her bones.

Maybe Camille harangued her for information because he knew more about the man than he'd let on and was waiting for her to confirm his suspicions. Then again, he could very well be tormenting her simply because he could. She remembered the way Suzette had cried after he'd paid his visits to her – a girl who'd put up with all sorts of things from all sorts of paying men. Madeleine thought too of his refusal to tell her what would happen if she couldn't get the information he wanted. She could guess the answer: he wouldn't give her the money, nor another job; she'd be back in the brothel for the rest of her days, poor little Émile with her. And the man might, just because he could, just because it was the sort of man he was, decide to do something to punish her.

'And find it something to eat,' Reinhart said.

'Yes, sir. I'll make up some paste for him.'

As she left the workshop, desperate to be alone for a few

moments, Madeleine noticed that Véronique had picked up a doll: part-made, its porcelain head sagging to one side, its face still a blank oval. She couldn't rightly say why it bothered her. It was a nagging, uneasy feeling, like a half-forgotten dream.

'My father's helping me to make her,' Véronique said, noticing Madeleine staring at the doll's face. 'The idea is that her eyes will move.'

Madeleine nodded, unsure what to say to this. What kind of child would want a toy whose eyes followed them? What kind of girl would make one? She'd thought at first that Véronique was as green as could be, but in some ways she was strangely knowing, and in others she was damned peculiar.

It was only when she was back in her room, removing her boots, that the memory, drowned long ago, resurfaced: a man with a white wig and a face wrinkled like a paper ball. The wig had stunk, she remembered – a vile nutty odour as though some animal, long dead, had once made its nest in there. He'd liked her to sit on his lap, that man, as if she were his granddaughter. And once, to 'reward' her, he'd brought her a doll, a bright new doll with a porcelain head. She felt her mouth flood with bile and for a moment had to sit down on her spindly chair, her head bent forward, her eyes closed.

Stupid girl. That was over long ago. It had been, what, ten years since? More? Their mother had sold their *pucelages* young, when she could get the highest price for them, then doused them in pucker water, and sold them all over again, and again. She'd had no choice, she'd said, though it had broken her heart, for their father was good for nothing, and it fell to her to keep them from the streets. Well, perhaps that was true, but it hadn't made it any easier, certainly not for little Suzette, whom she'd had to wash and dress before it was her time, and then again after, the bruises flowering on her arms.

Madeleine shook the thoughts from her mind. Why dredge this up now, years after she'd pressed it all down, months after Suzette had died? She needed to concentrate on the present, and on giving Émile a future. She needed to find something better, something clearer, something that established once and for all what kind of man Reinhart was.

When Madeleine went to the kitchen to begin the rest of her chores, she saw Joseph near the fireplace, cleaning his master's boots. For a moment she stood in the doorway, watching him polish the leather to a shine, his shirt pushed up past his elbows, showing the muscles beneath. He stopped his work and looked at her.

'Something is wrong?'

Many things, she might've told him, but instead she said: 'On the Pont Neuf there was a man selling leaflets, talking about a seamstress's daughter gone missing. So that's three gone, at least, isn't it? Maybe four.'

He nodded. 'I heard of the girl this morning from Nicol, one of the other valets. She was sent on an errand to the Rue Galande and never returned. Fifteen years old, he told me – a pretty girl. He says too that people are talking of man dressed in black snatching children from the streets.'

'What man?'

'I don't know. But apparently this is what some people have claimed: that there is this man who chases children.'

'What's he doing with them, then, this man?'

Joseph looked at her. 'I do not like to think, Madeleine.'

'No. Neither do I.' She was thinking, though, of two orphan sisters she'd once heard of, taken from the Marché des Enfants Rouges and auctioned off to the highest bidder. And, then, though she couldn't say why she felt such a need to tell him, Madeleine said abruptly, 'I saw my mother. In the street.'

'Ah. It must be good to have family nearby.'

'Yes.' She twisted her hands together. 'Your family – they're ...?'

'In Martinique.'

Martinique. It sounded beautiful. She had no idea where it was.

'Do you write to them?'

'I've paid for others to write to them, but no reply comes. Most likely my letters never even reach them.' He continued his polishing.

'Why?'

'Because they're slaves, Miss Madeleine. Martinique is an island of slaves now, cutting sugar for the French.'

Madeleine pulled at the fabric of her skirts, ashamed of her own ignorance. 'Will you ever go back there?'

'It may be that I'll have to. They've said that black men may not stay in France for longer than five years, but I've been here many years and so far no one has sent me back.'

'Do you miss them – your family?'

He shrugged. 'I miss something. It is difficult now to re-member them.' His eyes seemed to have lost their lustre and it occurred to Madeleine that his regular look of blankness was a studied expression, just as hers was – a mask to hide the pain.

'This is why I say it is good to have family near.'

Madeleine nodded. Would she miss Maman or Coraline if she was sent away, or if she never saw them again? It seemed to her she experienced little in the way of emotions these days. Even the loss she felt for Suzette was more of a dull ache than a stab, like a broken arm that'd been set bad and would now and again twinge with pain.

'Yes,' she said, going to the cupboard. 'Yes, I can see that.'

He was watching her. 'But then I suppose it depends on the family.'

She gave a half-mile without turning to look at him. He understood her too well. She'd have to watch herself.

★

Doctor Reinhart began work on the automaton box at once and in earnest. His moods, it seemed, shifted like quicksilver, for the anger of the morning was gone, replaced with a cold, almost mechanical drive. To Madeleine's growing anxiety, she was unable to properly search the workroom again either that night or the next, as Doctor Reinhart rarely left it, barely slept, but drew plan upon plan, creating detailed designs, Véronique working alongside him. Workmen had been brought in to make the springs and screws, cylinders and cams. A journeyman clockmaker had arrived to fulfil the orders Reinhart had no time to complete. Madeleine carried them trays of biscuits, fig tarts, little cakes encrusted with sugar, and saw Véronique handing her father tools and pieces of metal or working on her curious doll. Though she took in as much as she could on these brief visits, she saw no signs of new experiments, no clue as to the cadaver Reinhart was seeking, nothing by which to judge him at all. She knew her report for her third week here would need to contain something meaty if she was to keep Camille content and so far she only had scraps. As the days passed, the panic built in her chest and beat its tiny hammer in her head.

She'd hoped to speak more to Joseph, to try to find out whether he knew more about his master than he was letting on. He, after all, shaved the man every morning, helped him dress, shined his bleeding shoes; he must have information she could tease out of him. And besides that, she found she wanted simply to be near him, drawn like a moth to the light. She'd had only little time to speak to Joseph alone, however, for over the next few days he travelled across Paris carrying letters and orders and collecting items from various workshops: gold filigree from the Pont au Change, jewels from le Quai de la Feraille. 'Hacked from rock by slaves,' he told her, 'and brought across the oceans by boat.'

Véronique called Madeleine in to see the jewelled box when it arrived. Madeleine was stunned at first. She, the girl who'd only ever seen false pearls, paste and plaster, was now standing within a few inches of glittering gemstones – opaline teardrops, blood-red rubies – each one worth more than her and Coraline combined. Thinking that, she was flooded all at once with bitterness: that this aristocrat and people like her could order up the treasures of the earth merely on a whim, while she and most of Paris lived their days pressed down, the weight of fear upon them. For poverty wasn't just a lack of jewels, a lack of things, but a lack of power, control and safety. It was living life constantly on the edge, not knowing what tomorrow would bring. 'What will happen to me if you don't come back?' Émile had asked her when she went again to see him. 'Will she turn me out like she does the girls?'

'Of course not. And I will come back, *mon petit*. Don't talk like that. I'll come back a deal richer and things will be better, you'll see.' As soon as she'd said it, she'd cursed herself for a liar, no better than her own maman.

Véronique was watching her, she realised, must have noticed the downturned edges of her mouth. 'Do you not think it beautiful?'

'Oh, yes, it's lovely,' she said. 'The way the jewels reflect the light.'

'Then why are you frowning, Madeleine?'

She cursed herself inwardly. 'Keep your face smooth,' Maman had taught her. 'Never let 'em suspect you're not really enjoying it, or it'll ruin their pleasure likewise.' Perhaps Véronique suspected her of planning to steal the box. Of course, she'd considered it, turned it over like a piece of ice in her mouth, but she wasn't a fool. She'd be the obvious culprit, and that would be the end of her, neck snapped in the Place de Grève.

'It's only,' she said slowly, 'that it's an odd thing, isn't it, that there's all this here, sparkling, and yet so many who have nothing at all?' Whose lives are filled with darkness and violence; whose bellies are always empty.

Véronique returned the box to its velvet case. 'Yes, I suppose it is.' Then, after a pause, 'My father gives to the poor.'

The poor. As though they could all be lumped in one merry heap together: the needy, the wretched, the barely clothed, and then at the bottom, the entirely destitute – no food, no shelter, no shoes.

'Yes, Miss, I know. Tells us to give them food, too, if they come to the door.' It was an unusual act in this city where most looked through the homeless and limbless as if they weren't really there. And Reinhart hardly seemed to brim with love for his fellow man and woman. '*He makes toys for the rich and gives alms to the poor,*' Camille had said, '*but some say he does strange things.*' She wanted to ask Véronique about her father – about what he said to her when they were alone, about what she knew of his experiments, about why he observed her so closely all the time. But there seemed no way to ask without it seeming odd, and the more she observed the pair, the more she suspected that the man remained a mystery even to his own daughter.

'Do many come?' Véronique said.

'I'm sorry, Miss?'

'Do many people come asking for food?'

'Yes, Miss, quite a few.' Of course they did. They gave them the bones, the parings, the grease, for they'd eat anything.

'Children too?'

'Yes, sometimes children.' Often children.

Véronique nodded slowly, taking this in. 'And have you

heard any more, Madeleine, about those girls and boys going missing? I've been thinking about it.'

Madeleine hesitated. Why had Véronique been thinking about it? What relevance had it to a girl like her? 'Only rumours, really. Only servants talking about what might've happened to them.'

'And what do they say?'

'That there's a man in black clothing, stealing children, though I don't know where that's come from, or who the man's supposed to be.'

'Are the police looking into it? They must be.'

'They might be.' Not likely. The police, she might have told her, were only interested in the poor when they made a nuisance of themselves by begging or fighting or rioting for bread. Madeleine looked at Véronique with her unblemished white face; she wanted to ask her why she cared. For perhaps the first time since she'd been recruited, she felt a distinct uneasiness about her task – for wasn't it a rather shabby thing to do, to deceive a young girl, to fox a whole household? Necessary, of course, but still: how much simpler this would all have been had Véronique just been a selfish little rich girl with a heart as dull as stone.

Another three days went by and still Madeleine discovered nothing, for Reinhart kept up his frenetic pace, focused entirely on the machine. It was agreed that a friend of Joseph's, a young boy named Victor, would assist with the work of the house and the many orders. 'His slave master is a furrier who is most of the time drunk,' Joseph explained to them that evening as they sat playing at cards. 'He needed some extra work.'

'Extra food too, by the looks of him,' Edme grumbled, looking at the boy: huge eyes in a bright, thin face, a small silver hoop in one ear. 'Does your master not feed you?'

Victor shrugged. 'Not so often, Madame. Mostly he feeds his gullet. But I am stealthy, quiet like a mouse, and when he's in his cups, so I am at the grain.' He took up another jam tart and smiled at her. 'Nothing like this, though, Madam Cook. Here I am a king.'

Edme frowned at him, but her cheeks flushed crimson. She'd drunk it down like melted butter. As she picked up her cards, Madeleine watched the boy from the corner of her eye. He would, she thought, go far.

'How old are you, Victor?'

He shrugged. 'Eleven, maybe? I've been in Paris for at least two years.'

Then he was brought here when he was Émile's age. Madeleine thought suddenly of her nephew in his bed alone and her stomach curled in on itself. If she didn't get the answers Camille demanded within the next ten days, she'd have to return home empty-handed, with no hope for his future. She laid out her cards: king of hearts, queen of diamonds. And that was assuming that they let her go, that she could return to him at all.

After the boy had left and Edme had gone up to her room, Madeleine stayed sitting in the kitchen with Joseph to play at piquet.

'How old were you when you came over to France, Joseph?'

'Ten. Ten years old and stuffed in a sack to be sold.'

She stared at him, imagining the rasp of the sacking on skin, the terror of a ten-year-old child. 'Who in the name of Christ would put a young boy in a sack?'

'Many people would for the right amount of money.' He continued looking at his cards. 'It's a nice ornament, a slave boy, you see? Like a parrot, or a pet tiger. But boys, like parrots or tigers, realise they're supposed to be free. Particularly when they are beaten.'

'Did you try to run away?'

'Yes, aged twelve I ran away from my old master, but it didn't take long for the Watch to find me. And when they brought me back? Much, much worse. For a long time. Until Doctor Reinhart found me.'

'Found you where?'

'My old master was a trader of medicines. That is how Reinhart met him, and saw me. He bought me from him when I was fourteen years.'

Madeleine narrowed her eyes. 'Why?'

He glanced at her, caught her expression. 'Yes, I know Doctor Reinhart seems a strange man. I thought that at first too. I thought in fact that he would cook and eat me.' He smiled at her and her heart squeezed. 'But I have worked with him now for many years, and I think he is good within. He saw the marks on me, he saw what my master did, and he decided to stop him. Also, he was interested in me.'

'How d'you mean?'

'Come. I'll show you.' Joseph led Madeleine into the ante-room near the workroom in which the preparations of body parts and animals lived: the bloated embryo and the folded snake, the delicate skeleton of a bat. He walked over to a cupboard in the corner of the room and used a tiny key from the bunch in his pocket to unlock the dark wood door.

Heads. The cupboard was full of wax heads, perhaps twenty of them, some adults, some children, with the eye holes vacant. Joseph pointed at one, and Madeleine saw that it was him, his features reproduced in putty-colour, and she thought at once of the death masks they made after criminals had been hanged.

'Why on earth does he keep them?'

'For studying: what different people look like, different races, different ages. Look.' He pointed to the mask next to his. 'An

Indian, you see? And an aboriginal. So, Reinhart is interested in me because here I am different. That is what I mean.'

Madeleine realised her hand had gone involuntarily to her face, her fingers pressing at the scar. 'I see.'

Joseph made to close the cupboard, but before he did so, Madeleine noticed another face she recognised, the sockets empty. It was Véronique.

★

Madeleine woke to the sound of Edme rapping urgently on her door and dragged herself from bed. It was, she knew, the second of March, the day Madame de Marinière was due to return. It also meant she had only two days left to find the answers Camille wanted. For almost a week now she'd barely slept, barely eaten, consumed with growing fear. Despite all her searching and spying, she still had only fragments of the puzzle, and fragments were not enough. She'd written to Camille to tell him of the rows of masks, of the visit from the woman in silver. She'd told him of how Reinhart stayed up late into the nights, his workroom descended into chaos. 'There have been no more visitors, no more night-time visits, and nothing else I can tell you of. I've seen no sign of any strange experiments, though I've watched both night and day.'

No answer had come, no further questions, and the sound of his silence grew. She could imagine her mother, the rage rising off her, when she found that the daughter she'd long deemed worthless had once again come to nothing.

Bleary from lack of sleep, she collected the jug of water Edme had left outside her room and washed herself, using a rag to clean her face, throat, under her arms, between her legs. She wondered briefly how she washed, the mysterious Madame de Marinière. She imagined a troop of valets carrying hot water to her copper bath, a fog of exotic oils and perfumed soap. She'd heard that the rich bathed at least once every week, though there were still those who believed it dangerous. Madeleine dried and dressed, imagining for a moment that it wasn't yellowed petticoats and a coarse woollen dress that she pulled on, but a silk chemise, a diamond garter, a silver-threaded dress with an ermine stole.

'Madeleine!' It was Edme again. 'What on earth are you about? It's nearly six o'clock.'

She opened the door and Edme stood before her, her tired, puffy face like a lump of proving dough. All thoughts of diamonds dissolved and Madeleine hurried down to the front room to light the fire in the grate.

Doctor Reinhart didn't appear at breakfast, but remained closeted in the workshop, insisting that no one else should enter, even his daughter. Madeleine listened at the door to try to work out what he was up to but could hear only what sounded like the crackle of a fire, and a high-pitched sound, perhaps the cheep of the mechanical bird.

Véronique fidgeted at the table, leaving her bowl of *café au lait* to go cold. 'What if she isn't pleased with what we've made? What if she doesn't come?'

'She'll come,' Madeleine told her. It had been a test, she'd put a wager on it, and the woman would want to know the results. Just as Camille would arrive in two days' time to confirm that she had failed.

Sure enough, at eleven o'clock Madame de Marinière swept into the hallway, her hair glistening with jewelled butterflies, her skirts encrusted with silver. 'I trust the *cadeau* is ready?'

Doctor Reinhart, clutching a red velvet box, led the woman through to the parlour. He looked drained, Madeleine noticed, as though his work had leeched some of the life from him.

Véronique followed them into the room, Madeleine and Joseph remaining at the doorway, Victor behind them, straining to see. When Madame de Marinière was seated, Reinhart removed the brocaded lid to reveal what lay within: the dazzling oblong box of intricate silver filigree and enamel inlaid with pearls, diamonds, opals and emeralds. Doctor Reinhart opened the side of the box and removed a tiny key. This he inserted in the front and wound five times. He set the box down, pressed a catch on its side and all at once the top of the box sprang open to reveal a small silver bird beating its delicate wings. Madeleine held her breath. Though she'd seen many of the parts that composed Reinhart's creation, she couldn't quite believe the finished thing. As it rotated on its perch and sang, its shape, sound and motion were those of the real blue jay, but this bird glinted with topaz and sapphires, its eyes tiny chips of jet. It was, she thought, a bloody marvel.

The woman clapped her hands. 'Exquisite! Look at him! He's magical.'

And then she gave a gasp, just as Madeleine did, just as the servants in the doorway did, as the bird lifted from the box and flew, hovering only a few inches away from the woman's face, its wings a blur of white light. Madeleine held her breath, unable to believe what she was seeing. It was as though the little jay had been truly brought to life. After a few seconds the bird returned to its box and its song slowed, the lid closed. All was quiet save for the sound of the sound of the woman's silvery laughter.

Reinhart seemed delighted by the success of the bird box, more animated than she'd seen him before. After the woman's departure he went about his workroom whistling and tidying and then declared he was going out for dinner. Véronique too seemed happy, or at least relieved, that the project had been delivered. Madeleine, however, felt uneasy, nauseous, her chest fluttering like the bird's uncanny wings. How had it been possible to make it fly like that? Was it a trick of the eye – had it been attached to a hair-thin wire – or was there some mysterious sorcery? Her mind returned to the woman in the market with her brace of hares and her claims of dark powers. Only Victor, the boy, seemed to share her fears. 'Is it magic, Miss?' he whispered to her in the hallway. 'Is it some kind of witchcraft?'

Madeleine shook her head. 'You mustn't say that, Victor. Not here, not anywhere. I don't know how he does it.'

She pushed her fears to the back of her mind and went about her daily chores, washing, dusting, fetching. When she went in to clean the parlour, though, her heart seemed to stop. The blue jay was no longer sitting on his perch. He was lying on the bottom of his gilded cage, his bright eyes now glassy and blind. She opened his cage and put her hand on his soft feathers, felt his little body already cold; felt coldness spread through her own stomach. Surely it must be a coincidence.

She started at the sound of footsteps behind her and withdrew her hand from the cage. She turned to see Véronique, her eyes fixed on the dead blue jay, a mournful look on her face. '*Mordieu,* what a pity. That man must have sold us a sick bird, mustn't he? Or maybe it was just too old.'

'I don't think so, Miss. I reckon I'd have been able to tell.' When Véronique raised her eyebrows, she added: 'My father was a *oiseleur.*'

'Really?' Véronique looked at her sharply. 'You didn't say that when we were in the shop.'

'No, I didn't.' Because she hadn't wanted to talk about him, hadn't wanted to think about him, and because she'd assumed Véronique wouldn't care. But it seemed she'd been wrong about Véronique, in more than one way.

The girl's gaze had returned to the body of the bird, and her face had lapsed into a frown. Was she thinking what Madeleine was: that the timing was too peculiar? Could she be beginning to suspect, as Madeleine did, that Reinhart's powers went beyond those of a talented scientist and into a different realm?

But no, Madeleine chided herself, of course Véronique didn't think that. It was nonsense talk, just like the black magic that people whispered about in the streets to explain the coming of a sickness, or the failure of the corn crops; the disappearance of their children.

'It's probably,' she said, 'that his little heart gave out. Perhaps he had a shock.'

Yet something was tugging at her mind; something Véronique had said.

All afternoon, as Madeleine continued her work about the house, her thoughts continued to churn, blossoming into fears and theories, mangled monsters, half-formed. As she swirled the mop over the kitchen floor, Madeleine thought about the mechanical jay and how it'd come to life just as the real bird was dying. She thought of Franz and how his hutch had grown empty at the very point the machine was completed. Could Reinhart somehow be using the animals to give life to his own creations? Or was that the stuff of peasant superstition, a sign of her own ignorance?

And there was another fear spawning like a fungus in the darkness of her mind. '*Too old*,' Reinhart had said when rejecting the wares brought by the corpse-sellers. She'd assumed then that he meant too decayed, too rotten – but what if he hadn't meant that at all?

9

Véronique

Véronique knew that she must bury the bird. Descartes had said that animals had no souls, but she wasn't convinced that was right. She chose a patch of earth by a tree in the Place Dauphine. Madeleine dug with a little spade and Véronique lowered into it the box which served as a coffin for the early-expired jay. She threw some petals over the box, then Madeleine filled in the soil.

'There. A better ending than poor Franz's, at any rate.' Véronique stood up. 'No doubt you think me ridiculous, Madeleine.'

The maid give a non-committal shrug. She seemed tired, Véronique thought, hollows beneath her eyes, her skin a greyish hue. 'My little sister and I would often have little funerals when the animals in our father's shop died. Processions sometimes.' She gave a wry smile, brushed the soil off her skirts.

'I thought you said your sister was older than you?'

Madeleine turned away. 'No, I had another sister. A younger one. Died a few months back.'

Then I'll be like your little sister, Véronique had said to Madeleine when she first came to the house and her toes curled at the memory. 'I'm sorry. I hadn't realised.'

Madeleine shrugged again. 'No reason you should have.' She started back towards the house.

No, Véronique thought, as she followed behind her; no, there *was* no reason she should have deduced it, as Madeleine had never mentioned another sister, even though Véronique had asked about her siblings. Nor had she talked before today of how her father was a *oiseleur*, not even when they were standing in a *oiseleur*'s shop much like the one he must have owned. Her maid had grown thinner, she noticed, her collarbones more pronounced. Odd, given Edme's pastries.

'Madeleine,' she said, as they reached the door.

'Yes, Miss?'

'Is there something troubling you? Something wrong?'

Madeleine looked at her sharply. 'No.' A pause. 'I merely worry a little for my nephew, on his own.'

'How old is he?'

'Eight years.'

'Your mother looks after him, does she?'

She hesitated a fraction of a second. 'In a manner of speaking, yes. Only with the children going missing, I want to check on him, is all.'

'Well, you must go to see him this very afternoon. We've all been too busy these past weeks.'

Madeleine smiled cautiously. 'Thank you, Miss. I'll do just that, once I've spoken to Edme.'

As Véronique continued alone along the hallway towards the workroom, she thought back to the day Madeleine's mother had appeared on the bridge, running at them like a boat in full sail. Madeleine's manner had been stilted and nervous. No, it was more than that, Véronique thought, recalling the way they'd spoken. She pictured the woman's watery gaze,

Madeleine's sudden stiffening. It reminded her of Clémentine whenever she saw Soeur Cécile, her body rigid with fear.

'She was brought here as a child,' Clémentine told her. 'That's what the cook said. Brought here as a girl, like we were, and never allowed to leave.'

And instead of finding peace as some had, the woman had found only a bitterness that she turned both inwards, onto her deprived, self-scarred body, and outwards, onto those whose lives were younger, brighter, happier. 'What need has the body for honey or tenderness,' she would demand, 'when embraced by the sweetness of His love?' It was suffering that brought them closer to God, she said. That was why she hurt them.

Véronique knocked on her father's workroom door.

'Come in!'

Once inside, she saw that her father had opened the case to the wax anatomical doll and dragged her to the centre of the room. 'There you are. As we have missed some of your lessons this past fortnight, I thought we would embark on a course of intensive study, beginning with Violetta.'

It was his name for the anatomical girl, who'd been brought in a case from Italy. Véronique had seen her before, when she first returned to the house, but had never studied her at length. Though she'd seen the line that ran about the doll's middle, she still started with surprise when her father lifted off the stomach to reveal her internal organs.

'*Et voila!*' He began to remove the wax parts from the model: the lungs, the womb, the intricate heart, to show how each piece was constructed. 'Not as good as a real cadaver, but clever, isn't she? Beautifully done.'

'Very.' And she was indeed beautiful. Yet the model, so faithfully and accurately rendered, reminded Véronique of another girl's body, and though she tried to concentrate on the lesson that followed, her thoughts were ragged and disordered.

'And you see this, yes? The kidneys.'

As Véronique felt each piece, her father asked her questions about it: how did the blood flow to the heart? Where were the fallopian tubes? Did she know where the appendix was?

Véronique was acutely aware that he was testing her and finding her wanting, for she did not always respond at once, her mind being clouded with the past. After a time she felt a headache begin to burrow its way into her temples and when her father asked her yet another question she snapped, 'I am doing my best, you know.'

Reinhart stared at her. 'I do not doubt that, Véronique. I thought you wanted to learn my art, and there is a considerable amount to learn.'

'I do, but … Why did you not keep me at home?' she asked in a burst. 'Teach me yourself from when I was young?'

He stared at her, unblinking. 'I considered it. You were a clever child. Quiet, curious. But … the house was lonely without your mother in it and I am not, I think, very good with children.'

Véronique said nothing, but she didn't remember him being uneasy with her, not as a child. On the contrary, she remembered him as a kindly figure: the scientist who'd let her into his study, who had shown her how the magic was made. It was only now that she was older that the tension had arisen. Had he changed – grown more strange living here on his own – or was it she herself who'd grown awkward?

'I also had my own work to do,' he continued, seeming intent on justifying himself, 'my own practice to establish. Others told me that it was not right to keep a young girl here alone amidst the clocks in an empty house. They thought you needed the guidance of other women, and as I did not wish to marry again it seemed like the best option. The convent was recommended.' A silence stretched between them. 'And you are back now. We are improving your education. Making up for lost time.' He closed the lid on the doll, closing the conversation.

Up until then, Véronique had considered asking him what he intended for her, whether Lefèvre was right about her future, whether she was really to be his apprentice. But she knew now that it was not possible. She was not even sure she wanted the answer. Because in the recesses of her mind there grew a fear that he was training her up for a reason that was nothing to do with improving her education, but entirely to do with his own career. She pressed her fingers to her temples, trying to rub out the pain.

'You are tired, I think,' he said. 'We have all been working hard. Go now and get some rest. We will resume our lessons tomorrow.'

Back in her room, Véronique sat on her bed looking at the wooden-faced doll propped on the mantelpiece, the Christ doll that each of them had been given to tend to and to love. She felt suddenly terribly lonely. She took down the doll and held it on her lap as she used to do at the convent, but there was no comfort it in anymore; it made her feel no less alone.

She remembered how she and Clémentine had made beds for the dolls from handkerchiefs and old linen. 'I don't see why my one has to suffer. He's only a child, after all.'

And then all at once she didn't want to think of Clémentine anymore, or any of it. She didn't want to see the doll's limp, pathetic body. 'What did you do?' she asked it, 'when I prayed for help? You did precisely nothing.' And she threw the doll so that it hit the wall and slid down – broken, dead.

10

Madeleine

Hooves on stone, shoes on cobbles, the careful closing of a carriage door. Madeleine forced herself to climb out of her warm bed and into the night-time cool of her room. Peering between the curtains, she saw that it was the same carriage as before, gleaming oxblood in the carriage lamp. Once again, she crept down the stairs, feet bare and frozen, her pulse thudding deeply in her throat. The men, though, must've been standing on the far side of the room, as – though she pressed her ear to the crack – she could hear almost nothing of what was said, only the low murmur of voices. Then, all at once, footsteps growing louder as they approached the door where she stood. Madeleine had no time to run back up the stairs so instead flattened herself against the wall behind one of the grandfather clocks.

She waited there, her mouth flooded with the bitter taste of dread, listening to the sound of the men coming nearer. They stopped what must have been only a foot or so away, so close that she could hear their breathing just above the frantic beating of her own heart. She could smell the reek of death that seeped from their clothes and unwashed skin.

'We'll be in touch when we've anything else.' The man's voice was flat and colourless.

'Yes. This will be enough, however, for now.'

There was a long pause. Madeleine pressed her eyes shut and prayed. For a long, sickening moment, she was convinced that they'd heard her, that they were coming for her. But then they began to move away, their feet shuffling together along the passageway, and Madeleine finally breathed out. The smell of death still hung in the air.

The front door closed and then she heard Reinhart's slippered feet making their way up the stairs back to bed. Madeleine waited a minute or two longer until she was sure he was in his bedroom, then darted to the workroom and put her hand on the handle. It did not turn. She pushed harder. It was locked. *Merde.* She exhaled, trying to calm her breathing, then lowered her eye to the keyhole. Nothing but inky blackness.

Madeleine returned to her tiny room and huddled beneath the covers, but she knew sleep was a long way off. There was a child's body in that room, she was convinced of it now. Was this where the missing children of Paris were going: onto anatomists' tables? It would make sense. She'd seen those sort of men before – dead fish eyes, filthy hands; she'd heard of the things they were willing to do and she'd smelt their scent of death. Why bother digging in a stinking graveyard, when there was fresh flesh to be had for free in the streets?

When she closed her eyes she saw Suzette: dead – fair hair, blue lips – stretched out in a case, her arms crossed over her chest, her eyes staring glassily upwards. The waxwork doll, but real. Madeleine clutched at herself to stop herself from shivering and burrowed deeper beneath the sheets, trying to stop her mind from producing its flickering show of death. She tried to picture instead her sister's face from when she was alive, from when they'd been out walking in the Tuileries, Émile between them, sucking on a pineapple ice. But then the vision warped and she was back in the airless bloodied room with Suzette, the baby refusing to come out of her. 'Don't leave me, Madou. I'm so afraid.'

Madeleine didn't want to hear it. She pulled the covers over her head and waited in vain for sleep.

When she knocked on Véronique's door the following morning, her skin taut with tiredness, the coffee tray in her hand, Madeleine received no answer. She knocked harder, but still silence. She pushed open the door a crack.

Véronique's room was empty. The covers had been pushed back and her shift lay on the chair. She must have dressed herself already, but why? Unease snaking its way through her, Madeleine returned downstairs. As she reached the hallway, she heard murmuring from inside the workroom. The varnished door was ajar. Creeping closer, she heard Reinhart's voice.

'And this, you see, is what happens when you squeeze. You want to try?'

No answer but then, a moment later, surprised laughter. '*Mon dieu*, it truly works, Father.'

'Of course. And now if I cut here ... you see?'

'Yes.'

Madeleine stood frozen, horrified, her eyes round as copper pennies. They must be cutting up the body, the body of some poor child. And her mistress, instead of being repulsed, was merrily joining in.

She made herself walk into the room, her blood roaring in her ears, for she knew she must bear witness – must report it all back to Camille. She made her feet move her closer to the table so she could see the corpse laid out upon it. She could smell death now, a cloying sweetness that reminded her of the old man's wig. And then, as she drew nearer, she caught sight of the naked feet. She stopped, blinked, checked she was seeing right. For they were not, as she'd expected, the narrow feet of a child, but the splayed ugly toes of a man.

Reinhart looked up at her and in that moment – seeing both father and daughter together with their gloved and bloodied hands – Madeleine almost dropped the tray.

'Ah, Madeleine. You find us at work.' Reinhart looked positively cheerful.

'I was just wondering where my mistress was.'

'I dragged her from bed at the crack of dawn. It is always better, you see, to carry out a dissection in the morning light.'

Madeleine was suddenly overcome with the need to retch. She set the coffee tray on the table, put her hand to her mouth, turned on her heels and fled. As she left, she heard Reinhart saying, 'Well, really, this one barely smells. The man died only last night.'

After emptying her stomach in the yard, Madeleine walked slowly back to the kitchen. Stupid, stupid girl. Of course there was no dead child. She'd let her imagination run riot, through the charnel houses of her mind. The man merely wanted a fresh corpse in order to teach his daughter.

Edme, seeing her white face, raised her eyebrows. 'You know what they're up to, then.' She shook her head. 'I know, I know. Makes you sick to your stomach, doesn't it? Well, it's what anatomists do. I suppose he feels he has to teach her, like he would anyone else, but after all my years here I still don't like it – not one bit. It's not something for a young girl to see, even one who's grown up with his nonsense.'

Gaining no response, she peered into Madeleine's face. 'Sit down. *Sangdieu*. Have you not seen a dead body before?'

Oh, but she had. She'd seen her father's laid out on the kitchen flagstones, his hands clasped around his final bottle.

She'd seen her sister's young and bloated body, swimming in a sea of blood.

'I've seen bodies. Just not, you know, like that. Wasn't what I was expecting.'

Edme nodded, rubbed her shoulder. 'Best leave them to it. It'll be gone soon enough, and then we'll douse the place with vinegar. I'll make you up some ginger tea. I've found the stuff does wonders.'

Madeleine watched her as she boiled water in a pan. 'Is it always a grown one that he gets?' she asked after a while.

'What d'you mean, girl?'

'I mean – is it ever a child?'

'Lord, no. At least, not that I know of. I suppose I couldn't rule it out. They get them from the hospitals, mainly, or the prisons, or bodies left at the Basse-Geôle. Bodies that aren't claimed, you see.' Edme returned to boiling the water.

Madeleine doubted that body had come from those places, though. They hadn't waited for anyone to claim him – they'd carted him off when he was barely cold. That was something she could tell the police, but it wouldn't be enough.

★

The following day the jay's birdcage returned, but this time it was filled with sparrows.

'I caught them myself,' Victor told Madeleine proudly as he entered the kitchen, holding the cage up for her to survey, as the little birds flitted and cheeped.

'Very nice, but I'm not sure I understand.' She didn't like to put him down, but no cove was going to buy a sparrow. Unless he planned to paint them, of course. There were some who tried that trick.

'I used a stick and a laundry basket and a piece of string.'

'But *why*, Victor?'

'Because Doctor Reinhart asked me. Said he'd pay me three sous a bird.'

'Why does he want them?'

The boy set the cage down on the floor and crossed his arms. 'No idea. I didn't ask him. Perhaps to put in a pie.'

Madeleine shook her head. 'No, that won't be it.' The birds might well end up roasted on a stick, but they'd have a clear purpose before that.

'Do you, by any chance, have any pie, Demoiselle Madeleine? Or cake?' He gave her his best smile.

'You're a devil, you know that? A regular scoundrel. I'll go and see what's in the larder.'

As she stood up, she noticed a cut on Victor's eyebrow, the blood dried to a crust. 'Who did that, then?' she said softly. 'Your master?'

The boy looked down. 'I'm his property, aren't I? Like a dog or a spoon. He can do with me as he pleases.'

What could she say to that? Nothing at all, only cut him a thick slice of apricot cake and wish on his master a sudden death.

Madeleine didn't have to wait long to find out what Reinhart was about. She watched him take the cage of fluttering sparrows into his workroom ahead of Véronique's lesson that morning and positioned herself at the door so that she could hear what happened next. Within a few minutes of Véronique entering the room, she heard a thunk and then a gasp from her mistress. Peeking through the crack in the door, Madeleine saw something she'd never imagined she'd see: a bird's head, detached from its body, the beak still opening and closing.

Reinhart was crouched down next to it, a watch in his hand. 'Ten seconds,' he said triumphantly. 'The bird continues to move for a full ten seconds after the head has been severed.

The question is whether it is truly alive, or merely operating mechanically like our blue jay. At what point does death really occur? We will try another.'

Madeleine was so transfixed by what she'd witnessed that she didn't notice Joseph until he was right up beside her.

'What are they doing?' he whispered.

'Look for yourself,' she said, stepping back from the door. She heard again the thunk of the knife going down, the light pat as the bird dropped to the floor.

Joseph raised his eyebrows. 'This is a new one.'

'A new what?'

'A new life and death experiment.'

So this was it. Madeleine felt a mixture of relief and horror. These must be the unusual experiments that Reinhart conducted. This was what he had been about. 'Why does he do them?'

Joseph touched her shoulder and they both moved further away from the door. 'Because he is interested in life. Because he wants, I think, to understand how humans and animals work. Like taking apart a watch to see how it ticks, you see? He is forever trying things out, discussing with Lefèvre how it can be shown when life begins and ends, how chances of life can be improved. But this, the headless birds, this is a new game.' He smiled at her confusion. 'He is a scientist, Madeleine. He is a clever man. I know some of the things he does are strange, but you do not need to fear him and watch him, whatever you might have heard.'

'No, 'course not.' She looked away from him. It unnerved her how well he could read her. Did he know how much she'd been spying on Reinhart? Had he seen through her from the start?

It was the way that he looked at her too, as though he saw not something broken, but whole.

Madeleine's final report was not due until the following day, but she would waste no time. She wrote a note to Camille that night to tell him of what she'd seen over the past few days: the man's white body lain on the table, the cageful of cheeping sparrows, heads lopped into the floor.

'So you see, Monsieur, it seems that this is what he wanted the corpse for: to teach his daughter anatomy. Strange, yes, I grant you, but I'm told this is what anatomists do.'

She had to get it all down on paper, for herself as much as for him. She might, that way, cleanse herself of her other, stranger imaginings.

'It's possible,' she continued, 'that there's something else, but I think these are the experiments you'd heard about. The ones you said were unnatural. I thought you would want to know at once, Monsieur, so that you can tell your employer now.'

She'd done it, hadn't she? She'd got him his answers within the thirty days. The money would be hers, and then the freedom. She would get Émile away.

She put down her pen and sat back, exhausted. After so many weeks of strain and watching, she should have felt only relief, only excitement at the thought of a new life, away from her mother's house. But there was a thick trickle of unease seeping through it: the thought that there might be something else, something worse, hiding behind a locked door or buried beneath the ground. Isn't that what she'd sensed from the very first day, and never found the cause? She pushed it all away, blotted the paper and pressed in the seal of the fox.

At four o'clock the following day a boy appeared at the back door. His clothes were mere scraps of cloth and his face was as pale as bone.

'Is it food you want?' Madeleine asked. 'Let me see what I can give you.'

The boy was staring at her hair, at the strands that had escaped from her cap. 'Fox,' was all he said.

Madeleine stepped quickly into the passageway, pulling the door almost closed behind her. 'Who are you?' she whispered.

In answer, the boy held out a slip of paper.

She took the paper and unfolded it. '*Go to the Hôtel Particulier opposite the Church of Saint Roch on the Rue Honoré. Ask for Monsieur L'Epinasse. Leave at once.*' There was no name at the bottom, no seal, but she knew who it was from. She hid the paper in her dress, then handed the child a few sous. He pocketed them and ran. She walked slowly back into the kitchen where Edme was cutting up vegetables.

'Trouble?' she said, seeing Madeleine's face. 'You've gone pale as ghost.'

'My mother's taken sick,' she said. 'I need to go to her.'

Edme made a clucking noise with her tongue. 'It's this city. The dirt, the stench, the ill air. It's no wonder people are falling sick and disappearing *tous azimuts*.'

Madeleine looked at her as she deftly chopped carrots. 'I'd best go now,' she said, 'before supper. Mademoiselle Véronique is helping her father so shouldn't need me at any rate.'

Edme nodded. 'Go, then.' She pointed her chopping knife at Madeleine. 'But you'll not be long because I can't be doing the cooking and the serving at the same time.'

'Yes. I'll be back in no time.'

Madeleine ran to her room, stripped off her apron, washed her hands and face and tided her hair as best she could. Then, grabbing her cloak and hat, she left by the back door, the blood thundering in her ears.

With the rains of the past days the streets ran with filth, and though Madeleine tried to keep her skirts lifted out of the dirt, by the time she reached the Rue Honoré, her dress was spattered and her boots covered with the sticky black mud of Paris. She ran past the high-walled houses of financiers and merchants, past the workshops of fashionable dressmakers and milliners, their lighted windows glittering with gold-threaded fabrics and silver lace draped over porcelain mannequins. She imagined for a brief moment owning a shop of her own, making a living for herself. Capital. That was what you needed, her father used to say. A little bit of a capital and a little bit of luck, and surely she was owed a deal of both.

The *Hôtel* opposite the church was an elegant white-stone building with a cobbled courtyard across which Madeleine hurried. When she rang the bell, a tall man in dark blue livery opened the door and looked her up and down with disdain. 'Yes?'

Madeleine took the piece of paper from her dress. 'I received this note, telling me to come here at once. To ask for Monsieur L'Epinasse.'

He admitted her, then walked to a desk where another man in dark blue was working at his teeth with a toothpick.

The first man nodded towards Madeleine. 'Another *mouche*. For the Lieutenant-Général.'

Her heartbeat seemed to slow. Surely there was some mistake. They couldn't possibly mean him. She had no opportunity to ask, however, as the second man gestured for her to follow him, and then she was climbing a staircase of polished wood, leading to a carpeted hallway and then double doors, footmen on either side, who opened the doors and watched as she walked forward into a huge, high-ceilinged study.

At the other end of the grand room, behind a dark-wood desk, a man sat writing. He wore black robes and a long white, rolled wig. His face – fleshy, strong-nosed, dark-browed – was painted lead-white. Points of rouge stood out on his cheeks. Above him hung a gilt-framed portrait of the same man, but made slimmer, more beautiful, more powerful. Madeleine knew with a twist of sickness that it was him: Nicolas-René Berryer, the Lieutenant-General of Police, the most hated man in Paris. Another man sat to one side of the desk, his back to her. She could tell from his movements, his coat, his carefully brushed wig, that it was Suzette's least favourite client.

Madeleine stayed motionless, scarcely breathing, looking up at the little cupids that peeped out from the mouldings of the ceiling, the gilded swords and shields and scales of justice that gleamed from the panelling of the walls. Justice indeed, from the Paris police and a man such as Nicolas Berryer. She could hear the murmur of the men's voices from the other end of the room. The longer she stood there, the sicker she felt. She was painfully conscious of her unwashed hair, her mud-streaked clothes, her dirty and damaged face.

Eventually, Berryer looked at her. 'Come forward.'

As Madeleine moved towards the desk, Camille turned in his chair to face her. 'Monseigneur, this is Madeleine Chastel, the girl who's been spying on the clockmaker.'

'I know who she is.'

Madeleine felt a cold tendril of fear curl down her neck. She stood with her hands clasped, feeling the man's eyes crawl over her. 'Tell me exactly what you have found,' he said. 'I understand you wrote to Camille last night.'

Madeleine focused not on Berryer, but on the point behind him. Mechanically, she explained the experiment with the birds, the seemingly magical box, the visits from the stinking corpse-sellers, the dissection of the white-bodied man.

Berryer raised his eyebrows when she described the headless sparrows but looked otherwise unconcerned. 'Anatomists,' he muttered, 'they are a curious breed. But none of this is very unusual, nor unnatural. The flying bird must have been some clever trick you couldn't detect. What else?'

Madeleine hesitated. 'There isn't really anything else, Monsignor.' At least nothing she could clearly identify.

'These experiments on the sparrows, they're the only experiments you've seen?'

'Yes.'

Berryer and Camille looked at one another, seeming to agree on something.

'Who else comes to the house, girl?'

'Tradesmen, mostly. The men who create pieces for the clocks and automata. Occasionally a customer.'

'Yes, yes.' Berryer waved his hand. 'But anyone else? Any unusual friends? Who comes for supper?' He grimaced as Camille sneezed.

'He rarely has guests. There's a doctor, a Monsieur Lefèvre, who comes fairly regular.'

Berryer raised a charcoaled eyebrow. 'Oh yes. What do they discuss?'

Madeleine licked her lips. 'All sorts to do with science – electricity and blood-letting. That sort of thing. Mostly it seems quite dull. I've heard them discussing the soul. They talk about whether it's truly part of the body.'

'What else?'

'They talked of another man, Vaucanson. They said he'd failed.'

'The *mecanicien* who made the shitting duck,' Camille said to Berryer.

'Yes, I know full well who he is, man. It can hardly be

said he's failed.' Berryer looked back to her. 'Anything else? Anything about politics? Anything about the King?'

'I know Lefèvre gives the King lessons, but I've not heard them say much about it.'

'Well, what do they say?'

'That Lefèvre lets him ottomise dead animals, from the menagerie.'

'But you've heard of no plot, no sedition, nothing you thought untoward?'

'No.'

'You're quite sure?'

'Nothing that I've heard, and I have been listening whenever I can, but I can't always be everywhere.'

Berryer took out a small silver penknife and began to sharpen a quill. 'You say Reinhart rarely has anyone else to visit, few friends. He's an odd fish, then?'

'I'd say so. He seems mainly to stay in his workshop, inventing things.' Things no man should be able to create.

'He's been busy.' Berryer gave a slow smile. 'And what of the daughter?'

Madeleine's throat felt very dry. 'What of her, Monseigneur?'

'Does she have any strange habits, any unusual acquaintances?'

'No, I don't think so.' She was strange, yes, but she would not tell them that. She felt oddly protective of Véronique, reluctant to give them anything. 'She's only really interested in reading and learning, and in building her own machines.' She was, Madeleine thought, almost frighteningly ambitious and as sharp as Berryer's blade. She'd seen the girl's hands coated in gore, heard her laughing at a human heart. They'd been wrong that she was merely a naive young girl, but she wouldn't be telling them that.

Camille and Berryer looked at one another.

'Nothing in his correspondence?'

'Prolonged discussion about longitude and endless banalities,' Camille said. 'Plans, discussions about materials, that sort of thing. All unbearably boring. But it could, of course, be a code.'

Berryer turned back to look at Madeleine. 'What is your conclusion about this man? Does he seem to you to be dangerous?'

Madeleine's palms were sweating, her heart racing. How on earth was she to know for sure? She'd seen plenty of dangerous men in her time: men with glittering eyes and balled fists, men with a bellyful of liquor, a man with a poker in his hand; men with bitterness in their souls. She stared at Berryer. Reinhart was nothing like that, but it didn't mean he wasn't dangerous. Sometimes, it seemed, he was kind, but that in itself meant nothing – often it was the ones who at first seemed gentle that did for you in the end. 'As I say, sire, I've not seen anything that proves, exactly, that he's dangerous.' She hesitated. 'But I wouldn't say he was entirely a normal sort of man. He looks at me in a queer way sometimes, like he's measuring me up. And he does other odd things, things I don't like. Skinned a rabbit his daughter treated as a pet.'

'That's all? A dead rabbit?'

She wanted to explain that it was much more than that, that there was something unnatural in his nature. She wanted to tell them about the jay – how it'd died just as its mechanical double came to life. She wanted to tell them about the talk of bodies and her fear that he'd wanted a child's. But what evidence did she have to support these fears? None. Nothing more than a feeling lodged in her guts. And how could she say these things to him, the Lieutenant General of Police? He'd think her ridiculous and ignorant, little better than the crone from the market. 'There's nothing certain I can put my finger on, but I'm not sure that he's quite right, Monsieur. In the head.'

Berryer frowned at her. 'We are talking about a man who makes his living creating mechanical bats and pug dog clocks. Of course he's not quite right in the head. The question is whether he's dangerous. Have you seen anything aside from a skinned rabbit and a possibly illegal corpse to suggest he poses a risk?'

She hesitated, caught like a hare in Berryer's piercing gaze. 'No,' she said at length. And she knew as soon as she said it that it was a mistake; that they would hold her to it. That she might be wrong. But before she could add anything further, Berryer had turned to Camille.

'Well, then.'

'It goes forward now?' Camille asked.

'It is not up to me, ultimately, but I think we have enough for now to dispel what the last girl said.'

Madeleine went quite cold at that. Her face, in fact, must have drained of blood, as Camille, catching sight of her, smirked.

'You needn't look at me like that. Had I told you of the account, you'd have refused to enter the house, but as it turns out, it was clearly all invented.'

'What account, Monsieur?' she asked quietly.

Camille and Berryer exchanged a glance. The latter shrugged. 'No harm in telling her now, I suppose.'

Camille spoke easily, conversationally. 'There was a young girl who acted as a kitchen girl servant a while ago, only for a short time. She told the police that Reinhart had tried to do experiments on her.' He smiled.

'What experiments?' Madeleine's veins were flooded with ice.

'Oh, some nonsense about blood lettings and such. They

thought at the time it was far-fetched, and the other servants hadn't seen a thing, but of course it had to be ruled out.'

Madeleine did not answer. She remained rooted to the spot. Agathe had mentioned a kitchen girl her very first day in the clockmaker's house. 'More trouble than she was worth.' Because yes, that was trouble. That was plenty of trouble. Why had nobody told her?

'When was this?' she said stiffly.

'Some months ago, before your mistress returned. The girl had been reported for trying to sell a golden box she'd stolen from her master, and it was only at that point that she came up with this surprising story.'

Madeleine swallowed. 'Where is she now?'

Camille raised an eyebrow at her slowness. 'Madeleine, she stole a golden box from her employer. You know the penalty for that.'

Madeleine imagined the drop of the noose, the twitching of the body. For a moment she stood, not speaking until Berryer said, 'You can go, girl. Unless you have something further for us?'

She hesitated, the words forming in her mind, something the kitchen girl had made her think of; something she had to say. 'There was talk of an apprentice going missing from the Place Dauphine. The baker's boy. Then a Chandler's apprentice on the Rue de la Calandre.'

'Yes?'

'Those places – they're both on the Île de la Cité.'

Berryer's painted brow cracked into a frown. 'I don't see what that has to do with your assignment. Apprentices run away with unsurprising regularity.'

'Yes.' Again, she hesitated. 'But given what that servant girl said …' She breathed out, made herself continue: 'And other children have disappeared too, haven't they, from not far away? A seamstress's daughter—'

He sighed. 'This is women's talk, isn't it? You would do better to concentrate on the task we have set you than listen to idle tittle-tattle of idle domestics and low-bred whores.' He met her eye and she had no doubt then that he knew exactly who and what she was.

Berryer turned back to Camille. 'You were going to tell me about the coin counterfeiter.'

'Yes.' Camille glanced at her. 'Go back to the clockmaker, Madeleine. We're done with you here for now.'

Madeleine was trembling violently by the time she stepped back into the muddy street, so that when the woman approached, she didn't at first recognise her. Her hair lay lank upon her shoulders and her dress stretched over a swelling stomach. But it was Agathe, Doctor Reinhart's former maid.

'I need to get in there,' she said without greeting Madeleine.

Madeleine took in her strained, ashen face, the searching, bloodshot eyes. 'Agathe? What brings you here?'

'I need to see a police officer. Camille, his name is. I have to speak to him.'

Madeleine could only stare, fear corkscrewing through her chest.

'Please!' Agathe begged. 'Go back in there. Tell them I have to see him.'

'I can't.' She backed away. 'I was only here to report something and now I must get back to the house. I'm late. Why not ask them yourself?'

'They won't let me through the doors, they won't let me talk to him, but they might just listen to you.' Agathe clutched at Madeleine's arm, her fingers like a vice. 'I need to see him. Please. I've been waiting here for hours.'

'But why ...? Madeleine trailed off. It was obvious why she needed to see him. She could see the reason in her distended belly. 'That was why you left the house,' Madeleine murmured, almost to herself. 'That was the real reason you went.'

'He won't respond to my letters, and I'm all alone. My family won't take me back in.' Agathe's skin was yellow-white, her red-rimmed eyes were wild. Camille had done this to remove Agathe from Doctor Reinhart's house. So that Madeleine might enter. So that she might spy. Had Agathe refused to act as *mouche* herself, or had they not thought her right for it? Too faithful, maybe, and unable to read, so they'd ruined her life instead.

'They won't listen to me,' Madeleine said quietly, swallowing down her shame. 'Even if I could get back in there, my words would be so much straw.' And they'd damn her were she even to try.

'He told me he loved me,' Agathe whispered. 'That he'd set me up in an apartment.'

Madeleine stared at Agathe's pallid, bloated face. How could she have drunk in such an obvious lie? She felt part sympathy, part revulsion. She would never have been so stupid, but then she'd had a good education in distrust.

'If he won't see you, you're best off without him, I reckon.'

Agathe pulled her hand away from Madeleine's sleeve. 'All very well for you to say now you're safe and cosy in my job.'

Safe. Was she safe? She certainly didn't feel it.

'Agathe, what happened to the kitchen girl? Why didn't you tell me about her before?'

The woman frowned at her. 'Clothilde, you mean? She was a thief and a liar. She was only with us a month, and I've no wish to remember her.'

'She was hanged?'

'Of course. Many are hanged for less. Edme must've told

you, I suppose. I know she felt bad about the whole business, but the girl knew the law when she broke it.'

Madeleine tested the weight of her words, gauging how much she could risk. 'Do you think, Agathe, that Doctor Reinhart is a good man?'

Agathe's frown deepened. 'Good? What has that to do with anything? When did goodness ever help anyone in this city? He treated me fairer than the police, that's for sure.'

'But you never saw anything strange in that house? Anything queer?'

'Beyond moving metal creatures and snakes in jars?' She regarded Madeleine closely, assessing her, perhaps seeing her fear. 'I saw plenty, but since you won't help me, I won't help you.'

Madeleine went quite cold at that. Were there really things that she herself had missed or was this just a taunt to make her as afraid as the maid that she'd displaced? 'Agathe, please tell me what you mean. I'd help if I could.'

'Oh, you could help, but you're choosing not to. Well, I'll try again myself.' Her voice was brittle. 'I'll make him see me now.'

Madeleine walked after her for a moment, feeling guilt worm its way into her. 'You could go to the Hôtel Dieu, couldn't you? They'd take you in.' She knew as she spoke that it was hopeless.

Agathe gave a laugh, dry as a breaking stick. 'Hôtel Dieu! God's place in Paris. Have you seen that place? There's no God there, only the insane and sick, lying heaped in squalor. Like I said, he promised me. I'll make him remember his promise.'

Madeleine watched the woman move heavily away, watched her argue in vain with the guard at the door. Though it wasn't a cold day, she shivered. If Camille – if the police – would casually crush one woman to plant another in her place, then

she'd no doubt they'd trample her too if it turned out she'd
got it wrong; if it turned out she'd failed to warn them that her
master was not just unusual, but insane.

★

She only really understood when the summons came two days
later. Madeleine was on the stairs to the hallway when Joseph
opened the door to two men dressed in splendid red velvet
jackets with gold braiding. 'We must see Doctor Reinhart,'
one of the men announced. 'Immediately, if you please.'

Joseph flashed a wry look at Madeleine and led them through
to the parlour, leaving her to fetch Doctor Reinhart from his
workshop. She found him with Véronique and Monsieur
Lefèvre, who was showing them some kind of device from
which sparks of bright light flew.

'Can you not see, girl, that we are occupied?' Reinhart
demanded.

'I'm sorry, Monsieur, but I think you'd best come at once.
There are two men come, very eager, it seems, to see you.'

Reinhart sighed. 'Claude, you must excuse me, it seems.'

'No matter. We will continue another time. But I tell you,
this is the future.'

Grumbling about unwanted visitors, Doctor Reinhart made
his way to the parlour, Véronique and Lefèvre behind him.
The men in red bowed in perfect unison and one flourished a
letter, which bore a sign of three *fleur de lis*. 'Doctor Reinhart,
it is my great honour and duty to present this to you.' He went
down on one knee, like some devilish suitor, holding the letter
up high.

Reinhart, frowning, took the letter from him, broke the red
wax seal and unfolded the thick piece of paper that formed
both letter and envelope. As his eyes ran over the text, he

136

scratched his head beneath his wig. 'I think some mistake.' He sat down abruptly on a wing-backed chair and Véronique hurried over to him.

Lefèvre took the letter from his hand and read it aloud: '*On the orders of Argenson, on the recommendation of the Marquise de Pompadour, Doctor Maximilian Reinhart is appointed Horloger du Roi with immediate effect. He is to move at once into the Galeries des Louvre. His family and staff to accompany him.*'

For a moment nobody spoke, merely stared at the letter that Reinhart still held out before him.

Lefèvre himself frowned in incomprehension, then let out a great guffaw. 'Clockmaker to the King! Have they entirely taken leave of their senses?'

'I do not understand. I did not apply.'

The men in red stood awkwardly, glancing at each other, having clearly expected a rather different response. Lefèvre gave another chuckle that was somehow less convincing. 'You don't apply for positions in the royal household, Reinhart, you're simply chosen. And for some reason they have chosen you.'

Madeleine's eyes met Véronique's. This must be why she'd been tasked with spying on him, then: to judge if he was fit to enter the royal household, level enough to be clockmaker to the King. And his appointment, that must mean that she'd succeeded – that the police had taken her word. She'd be paid, then, wouldn't she? She'd be set free of Maman. She'd be out of the Rue Thévenot! Maybe they'd leave Paris and start afresh, someplace where no one knew what she'd been before, and where she might herself forget.

Lefèvre was ushering the men in red back into the hallway, saying that he wished to speak to his friend alone.

'Well, what will it mean, exactly?' Reinhart asked when he returned.

'It will mean, *mon ami*, that you will go to live in the Galeries des Louvre, alongside other selected craftsmen. It is a great honour, of course, though the rooms are terribly shabby.'

'And then?'

Lefèvre shrugged. 'And then no doubt Louis will come wandering about your workshop expecting you to explain things to him, demanding that you construct machines for his beloved chickens and such. You will have to learn some manners, man, or he'll have your head on a spike.' He looked again at the letter. 'On the recommendation of Pompadour, indeed. Well, clearly you have impressed someone with your clever little toys.'

'The woman who visited us,' Véronique said, inspecting the letter, 'Madame de Marinière. Was that her, then? The King's mistress?'

Madeleine thought of the woman in the blue and silver encrusted dress, the arched brows, the practised laugh, the hard glitter of a diamond. It was for her that she'd been working, she was sure of it. That was why the woman had stared at her that day, why she'd given her that curious smile.

'Yes, Miss,' Madeleine said. 'I think it must have been. Madame de Marinière: it was a joke, you see. Her real name is Jeanne Antoinette Poisson.'

Maman and Coraline had sometimes talked of Madame Poisson – how she'd risen from humble beginnings to become Madame d'Étiolles, then the Marquise de Pompadour, the most powerful courtesan of them all. But they'd talked of her as a woman like themselves, grown powerful and rich; a woman who was ultimately a whore. The woman she'd seen in the shop was nothing like them; she was a woman who knew her own worth and made her own plans, a woman worth a thousand of her mother. Thinking of that, Madeleine was seized with a watery sort of hope. For if Pompadour was her employer, then didn't that mean she herself was worth something? That she might be caught in the reflected shine?

'Yes, indeed,' Lefèvre said. 'It will be Pompadour who did the background work to find you, Reinhart. The woman never stops, it seems. Clearly the automaton box was part of your trial, so Véronique here did very well to accept the commission despite the, let us say, "ambitious" timeframe.' He winked at Véronique and she smiled at him in return.

Reinhart noticed none of this. He was staring straight ahead. 'Well, we have passed her test,' he muttered. 'We will have to continue to surprise.'

Madeleine emerged from the parlour as though in a dream, the pieces still melding themselves together in her mind. She'd accomplished her mission and so could soon slip away, claiming a sick relative, perhaps, or responsibilities to Émile, and of course the latter was true. She'd take him out of Paris, without discussing it with Maman – that was the only way to get clear. She felt a tug of sadness at the thought of leaving Véronique and Joseph, Victor too, even Edme, but she wouldn't be sorry to leave this house, or to say goodbye to her past. It was as though the weight that'd been pressing down onto her shoulders was finally letting up; for the first time in her life she could look to the future without it seeming closed in and dark.

Snapping out of her reverie, she noticed that one of the men in red was staring at her intently. At first, she thought he was just gaping at her damaged face, but then she realised he was trying to convey something to her, to approach him, perhaps. Why? When she thought herself unobserved, she stepped closer to him and his voice was a harsh whisper.

'The message I was told to give you is this: you are to stay in place. You are to assume nothing. Your job here is not yet done.'

Two

The Louvre

I I

Véronique

From a distance, the Louvre looked exactly as it should: a great stone palace, its grand colonnade as imposing as a rank of soldiers. Closer up, however, you could see the ramshackle wooden structures that tradesmen and salesmen had erected in the courtyard, the dirt that dulled the marble. Once inside, you realised you'd entered not a palace, but a giant hive, the residence not of royalty, but workers. The floors of the building had been subdivided both horizontally and vertically into numerous segments, some vast, some tiny, housing portraitists, poets, sculptors, scientists, with their underlings buzzing about them. Her father's workshop was a large, shadowy room on the second floor filled with dusty wall hangings and dark oil paintings in heavy gilt-wood frames. The rest of their apartments comprised two bedrooms, a library, a study, and a salon lined with tarnished silver mirrors, its ceiling covered with a faded fresco of Christ surrounded by nymphs. Into these rooms were moved the army of clocks and the menagerie of mechanical animals, joined by a stray calico cat who decided she'd make this her home.

When they'd first arrived here, Véronique had stretched out

on the salon floor, arms spread wide, shoes kicked off, and gazed up at the flying angels. How long would she be able to stay here? A month? Three months? A year? Might she dare to hope her father's appointment would bode well for her – might mean he would devote more time to teaching her – or would he want her out of the way? Did he intend something else for her entirely? Madeleine had found her there, and, when she'd taken Véronique's hand to help her up, Véronique had pulled her down, laughing, and insisted that her maid lie next to her and stare up at the ceiling too. For a while they had both lain there, side by side, gazing up in silence. It reminded Véronique of when she'd lain in the bed next to Clémentine's, both of them staring at the smoke-stained ceiling, whispering about what they'd be. 'We make our own future,' Clémentine had told her, but how could that be true?

'Do you think, Madeleine, that people can determine their own destinies?'

'How d'you mean?'

'I mean: can we really control what happens to us, or is the path already set out for us, from the moment we're born?'

'I couldn't say, Miss. It'd be nice to think we could make our own choices, but for many of the people I see their paths are pretty narrow and short.'

'Yes.' All those huddled, stunted people she saw in the streets, their faces prematurely aged. Could they claw out of the poverty they'd been born into? Unlikely. Not in the Paris she was coming to know; the Paris where children could go missing from the streets and the authorities scarce blink an eyelid. Madeleine said there had been further pamphlets, but she'd scoured the pages of her father's newspaper and the children got no mention at all.

'I've been given a chance, though, Madeleine – a chance at life. A chance to do what I've always wanted to do: to make machines of my very own. I can't let it slip away. I can't go back to the convent where I've spent most my life.'

'You won't have to, will you?'

'I don't know.'

A pause. 'It was bad there?'

'Yes.' Véronique saw Clémentine's face; the eyes of the statue, bleeding. The images were there and then gone again.

Madeleine didn't ask any other questions, but simply took her hand.

'So you see,' Véronique said quietly, turning her head to look at Madeleine, 'I must succeed. I must convince my father I should stay; that he should continue to train me. I can't go back there.'

'I understand.'

'You do?'

'Yes. I can't go back either.'

Véronique was sure, then. Something terrible had happened to Madeleine. Something of which she would never speak, just as Véronique would never speak of what she herself had seen and done. They'd both been formed out of darkness. They'd both had to construct themselves.

Their solitude was brief. Almost at once they were besieged by clockmakers and artisans who made their way through the enormous hall of the Louvre and up the staircase to congratulate and pay their respects to her father; fresh-faced courtiers who thought they might befriend this new appointee so as to bring them nearer to the King.

They failed, of course, all of them. Her father dealt with these people in his usual brusque manner, not wishing to be distracted from his work, but this didn't seem to dissuade them. There was always some beribboned hopeful hovering about the place, clutching a carefully made clock that he wished to show to the *Horloger du Roi*.

'Are we to feed them?' Edme asked plaintively. 'They come to the kitchens demanding *café au lait* and biscuits.'

'No, no,' her father assured her. 'It will only encourage them. If they ask, they may have bitter coffee and black bread. That is all. We are not a hotel.'

Faced with Reinhart's silence and Edme's coffee, they would try Véronique instead, asking her questions and paying her compliments as she walked from room to room in between her lessons. She couldn't deny that, at least at first, she liked the attention from these well-spoken and sometimes handsome men in pink satin breeches, red-heeled shoes and embroidered waistcoats of cinnamon, plum and peacock blue, like a flock of preening birds. After ten years in a convent, it was quite something to be noticed and admired – to be given gifts, to be made the subject of verses, even if their quality was a little wanting. (One courtier wrote an entire sonnet comparing her to a pedestal-clock.) It was not, however, her intelligence they claimed to admire, but her 'emerald eyes', her 'narrow waist', her 'skin of cream and honey'.

Walking back to their rooms with Madeleine one afternoon, Véronique passed two wine-flushed young men playing dice at a little table, their voices too loud in the echoing hall. 'Pretty, yes, but entirely deluded if she truly thinks herself his apprentice. Why train up a seventeen-year-old girl when you can have your pick of all the journeymen clockmakers in Paris? That's not why the old man keeps her near.'

She continued past them, her cheeks blazing, and as she climbed the staircase, the men's laughter ringing in her ears, she could feel the anxiety crawling through her veins, feel her chest tighten. For they had voiced her own fears. Why train her when her father could choose from the best? Why waste his time on a *girl*?

Back in her dressing room of pale green silk, she stared into the mirror, watching tears of anger form and shiver over her eyes. She barely paid attention as Madeleine began to unbutton her new silver robe, the dress that had at first seemed such a joy. From the hall below she could hear whoops of laughter, chatter, and she imagined they were all mocking her: the silly little girl who'd somehow imagined she'd be allowed to play with the boys. She wished they would wither and die. How long, she wondered, until her father called her in and told her it was time to pack up her things and make her return to the convent, this time as a novice, telling her, just as he'd told her at seven, that this was no place for a child? She saw again the long stone corridors, hurrying through the cloister, fearing what she would find.

'You mustn't listen to those boys. You're far cleverer and more talented than them. They just can't bring themselves to admit it.'

Startled out of her thoughts, Véronique lifted her eyes to look at Madeleine, reflected in the imperfect glass. 'That isn't ...' She swallowed. 'My father's never said that.' On the contrary, he'd said he was testing her, and she'd no idea if she'd pass.

'What *has* he said, then?'

'Very little. I've meant again and again to ask what he intends for me, why he brought me back here, but I find I can't. You must think me foolish to be afraid of my own father.'

'Oh no,' she said quietly. 'I don't think you foolish.'

Véronique watched Madeleine hang up the dress. 'What about your father, Madeleine? You've never spoken of him. Does he talk to you?'

Madeleine smiled. 'I'd worry if he did, Miss. He's been

underground near twelve years. But yes, he used to talk to me about all sorts. He was a good man. Weak, maybe, but kind.'

'How did he die?'

'As he lived: soaked in brandy.'

She said this matter-of-factly, as though it were the usual thing to happen to one's father, but Véronique understood now that Madeleine had several layers, and this was only the topmost one.

'And you were ...?'

'Eleven. Old enough.'

Old enough for what? 'Was that when you became a servant?'

'No, that wasn't until later,' Madeleine said. 'When it was the only option left.'

What other options had been run through, Véronique wondered. She considered again the scar on Madeleine's face and the story it might tell. She considered too her own options: was there some way to become an automaton-maker even if her father wouldn't teach her? If sent to the convent, could she run? But she couldn't bring herself to ask this woman whose own narrow path in life had been chosen for her.

'You remember him, at least,' Véronique said. 'I don't remember my mother at all.'

Madeleine pulled at the laces. 'You were a babe, weren't you?'

'Yes. I killed her.' She didn't know why she was talking like this.

'Nonsense. You were just born and she died.'

'No, I wasn't born, that was the problem. I wouldn't come out. In the end, Father had to cut me out with his scalpel.' She envisaged, as she had so many times, a room full of screams, a bed full of blood. 'He had to kill her to save me.'

Madeleine's tone, when she spoke, was surprisingly gentle.

'Well, otherwise you'd both have died, wouldn't you, as so many others do.'

'Yes, probably. I was breech – the wrong way around. So you see, I've always been awkward, always been strange.' She smiled and wiped away the solitary tear that had trickled its way down her face.

Madeleine was still at work on her corset. 'Better to be awkward than a sop, I say. Better to be sharp than soft, or others will take advantage.'

She gave a final tug to release the corset and Véronique felt a wave of relief. She understood then why Madeleine had seemed a closed box so much of the time. It was safer that way – to shut yourself off from other humans in the hope that they couldn't hurt you. But what kind of life would that be? As lonely and cold as a convent cell, or as one of her father's metal creatures.

'I tell you, you ignore those young men,' Madeleine said, hanging up the corset, 'they were clearly a pair of fools. If they're still there when I go back downstairs, I'll have Joseph throw them out.'

★

The following day Monsieur Lefèvre came, and he brought a visitor. A visitor with a retinue. 'Le Count de Saint Germain', she heard Lefèvre tell Joseph, who led them through to the salon to meet her father. Shortly afterwards, Joseph arrived to inform her she too had been asked to attend. He seemed unusually nervous.

Véronique knew who the man really was as soon as she entered the room. She'd seen his broad handsome face and prominent nose in portraits, in books, in newspapers, on coins. He wore a deep blue velvet jacket with buttons of gold. His eyes were large, rich brown, guileful. He was the man from the paintings, but older, fleshier, his lips swollen as though stung. It was the King.

'This is my daughter, Monseigneur. Véronique. I regret I have not yet had the opportunity to present her at court.' Reinhart stood to introduce her, and, Véronique, panicked, gave a low bow, such as the nuns had taught her long ago, and which she'd never truly mastered.

When she righted herself, she saw that Louis was regarding her with a half-smile. 'Your father is teaching you, I understand, Demoiselle Reinhart.'

Her heart was in her mouth. 'He is, Majesté.'

'I've been trying to convince Reinhart to let me teach her,' Lefèvre said. 'After all, I have a far more current understanding of anatomy. But he is selfishly keeping her to himself.'

The King's eyes were still on her, eating her up as though she were a morsel of cake, and she felt sweat beading in her hairline.

'Well, you must have lessons with Lefèvre. He knows everything about the body that there is to know; understands the pulse of life.' Still that half-smile on his powdered face, as if he saw right through her dress and shift to the flesh and muscle beneath.

Véronique gave another, smaller, curtsey, aware that her cheeks must now be crimson. 'Yes, sire. Thank you.'

He turned his attention to her father. 'And you, Doctor, you studied at the Jardin Royal, I understand? Then as a clock-maker?'

'That's correct. I began my education as a mechanician in Neuchatel, when I was a young man, under the tutelage of a clockmaker. It was he who advised me to come to France to study anatomy, once he realised my proclivities.'

Madeleine entered the room carrying a silver tray containing a dish of sugar plums and a pot of coffee. As she set down the tray, Véronique saw that her hands were shaking. She too knew the true identity of their guest. For an instant the girls' eyes met, astonishment passing between them. For an awful moment, Véronique thought she might laugh.

'I'd be interested to learn more about your studies,' the King said, 'and your creations.' Véronique noticed that his gaze was now on Madeleine, on her narrow waist, her small breasts, on the strange scar to her face. Was this what a king was supposed to be like? Having seen Pompadour's elegance, she had assumed the King would also be a person of refinement, intelligence and subtlety. The man, however, exuded a brutish sexuality, a dangerous sort of power.

'Lefèvre and I have been discussing how we may understand the anatomy by reproducing it.'

'Indeed, Majesté,' Reinhart agreed. 'There is no better means of learning.'

'I'm particularly interested, you see, in reanimation – in restoring life. I've heard much of your inventions,' the King said. 'I wish to see them. I confess few things have brought me such delight as the digesting duck that Vaucanson produced some years ago.'

Reinhart scowled and Lefèvre laughed. 'The defecating duck. The glory of France. You have him making useful machines now, have you not, sire?'

'Yes. Looms, and certain other important inventions.' Louis paused and was suddenly serious. 'In fact, that leads on to what I wanted to talk to you about, Doctor Reinhart. Perhaps you could show me some of your inventions and I'll tell you what I had in mind. It is ... delicate.'

Reinhart bowed. 'Of course, sire. Follow me.' He led the King and Lefèvre out of the salon. He did not invite Véronique to join them. Before he reached the door, however, the King turned back and smiled at her. 'I hope we shall meet again, Demoiselle. In fact, I'm sure we shall.'

Véronique's father didn't tell her what he'd discussed with the King, but whatever it was, it provoked a peculiar heating of his blood. As soon as Louis had left, he began to take down books from his shelves, and scribbled a letter that he gave to Joseph to deliver at once, instructing him to wait for a reply.

Then he and Lefèvre shut themselves away in the workroom, insisting that all callers be refused. 'Don't take it to heart,' Lefèvre told her. 'It's only that the King has commanded it remain entirely secret.'

Reinhart stayed in his workshop long after Lefèvre had gone home and the lights everywhere else in the Louvre had been extinguished. Over the next two days, he insisted on taking his meals in there, picking at the pies and roast chickens that Edme brought up to him while continuing his work. Joseph was sent out with further letters, marked 'confidential and urgent', and with orders for the tool-maker, the doll-maker, the moulder of brass. The boy, Victor, was brought back in to assist with the gathering of parts: tubes, wool, wax and wood, a variety of animal bones. Véronique was left to eat alone at the long table in the blue-walled dining room, the calico cat her only company, the scrape of her cutlery and the rustle of rats the only sounds in the silence. She began to find the fusty and ill-lit apartments as oppressive as she'd once found the convent. She felt unwanted and out of place. Was her education at an end? And what was it that was so terribly important, this commission her father had begun? What could the King have asked him to do that justified him forsaking everything else?

On the third evening, she knocked on the door of her father's workshop. He did not admit her, but stood in the doorway, blocking her way, his spectacles pushed down over his nose.

'You must understand,' he told her, 'that there is important work for me to do, work which may advance science. The timing is not ideal, but the thing itself will be.'

'Why can't I be part of it, Father?'

He scratched his head. 'Because the project is entirely confidential, and because … you are not yet ready.'

'I learn quickly.'

'Véronique, you are inexperienced. You have been much sheltered. Some people will frown upon what we are doing, what I have been asked to do.'

'But if the King himself—'

'Many disapprove of what the King does, Véronique. Many consider him immoral, a *libertin*.'

She thought of some of the things Madeleine had told her: how he'd taken three sisters as his mistresses, one after another, or perhaps two at a time. She thought of the way he'd looked at her so that her skin began to crawl.

'And you, Father – do you think it wrong?'

'On the contrary, I think it an advancement, but then I've never been squeamish.'

'Neither have I.'

'I know that, Véronique. But if it's known you helped us with this project, people might think you perverse, unnatural.'

And wasn't she? 'Will they not think that of you?'

'Perhaps, yes. But I am a man. An established *mecanicien*. You are a young demoiselle.'

'But that is not—'

'Véronique, I tell you that you are not *ready*.'

Not ready for what?

'This is the most difficult commission of my career. I need you to give me time. You understand?' His voice was firm, his eyes very dark and still.

Véronique stared at her father, at his silver-streaked hair

153

and chalk-white face, at the crumpled black suit, with its little buttons, shiny as jet. She understood. She understood all too well. The courtiers in the corridor had been right.

An empty sort of fury grew within Véronique then, as to her life, her future, her powerlessness. She knew herself to be gifted, to be as clever as any boy, capable of many things, yet she was being shut out just as she'd been shut in a convent most of her life, and not even told what it was they were making.

Without asking her father or telling Madeleine, she dressed herself in her grey travelling cape and a straw bonnet so as to look like any shopgirl or *grisette*, left the Louvre and strode out, beyond their district and into the streets and squares around. 'You must understand life,' he had said. 'You must learn from it.' Very well, then. Learn from it she would. She left the main thoroughfares and walked down the narrow alleys north of the Louvre where the sounds of printing presses and black-smiths' tools rang out, where ramshackle timber houses leaned towards each other, almost obscuring the sun. She inhaled the stench of excrement and death and unwashed bodies, she felt flaking paint and rusted doors and the softness of rotting matter beneath her feet. She saw women sitting in doorways, shelling peas or sewing; old men with crumpled faces sucking on earthenware pipes or simply staring. Occasionally, she saw barefoot children rolling wooden hoops along the street, and she began to recognise that taut, pinched look that made them seem older than their years: the unmistakable mark of hunger, of bellies never full. All her life she'd been shielded from the poverty and ugliness that lived here, and yet it existed almost on her doorstep. She felt ashamed of her own ignorance, and of her own fear.

'Inexperienced,' he had called her. And who had kept her shut up all her life, along with all those other girls, intended to be kept ignorant and placid? Oh, if she were to tell him what it had really been like there, how they'd treated the daughters of the *haute bourgeoisie* when their well-paying parents weren't looking.

Though there were some boys and girls in the streets, Véronique noticed that most went about in groups, or with their mothers or fathers. The people wore a frightened, hunted look. On a mildewed wall she saw a poster, freshly pasted: 'KEEP YOUR CHILDREN NEAR'. And perhaps because of the fear that she sensed in the city, or perhaps because of the approaching dusk, Véronique began to worry that she herself was followed, watched, through dark alleys and unpaved streets. Attempting to throw off her shadow, she turned back on herself and hurried along a narrow stinking passage, then along a series of winding, unlit streets in the direction, she hoped, of home. After a time, however, she realised she had no idea where she was. The houses were so high she could see no landmark she might have recognised, and the light was giving out. 'Breathe, Véronique,' she told herself, but panic clawed at her insides.

Hearing loud voices from not far away, she decided to follow the sound, thinking perhaps it was street traders or performers on a main street from which she could find her way back. As she drew closer, however, she realised that these were no street cries or actors but howls of real distress.

'They did nothing!' she heard a woman cry, and saw a knot of people standing about her. 'I begged them and pleaded with them, and they just shrugged their shoulders, said it was my fault for leaving him on his own. Well, what choice did I have? I had to work, didn't I? What else could I have done?'

Two other women were holding onto her, trying to comfort her, as others gathered around. 'Let me talk to them,' said a man with a face like a parboiled ham. 'I'll make those lazy *rousses* do their bloody jobs. Have they not heard about the other children taken? Are they deaf as well as stupid?'

'They don't listen because they don't want to listen,' one of the women standing closest to the mother hissed. 'There's a reason they're not looking for the children.'

Murmurings of agreement from people in the growing crowd. 'It's because it's an aristocrat that's taking them!' another woman said loudly. 'An aristo who bathes in blood. And what do the police care for the poor?'

Rumblings of approval.

'Who's to say it isn't the police taking them themselves?' another said. 'Who's to say the police aren't bringing them to the aristos, and taking their reward in gold?'

More shouting, jeering; more people, drawn by the noise, gathering in the filthy road. Standing in the shadows, her hood drawn up, Véronique listened to the people's theories: men with a thirst for children's blood, women who drank it to make them younger, witches who composed their potions from babies and bats and bones.

Frightened by their hostility, which seemed almost to shiver in the air, Véronique moved away from the crowd of people, to the square at the end of the street. Then at last she could see the shape of the Notre Dame. She could work out her way home.

'They were furious,' she told Madeleine when she returned to the Louvre, shaking with fear and cold. 'They were saying it was people of the nobility, bathing in the children's blood!'

Madeleine was sponging the black mud from Véronique's skirts, apparently intent upon her task. The calico cat was lying in the corner, washing her tri-colour coat. After a long moment, Madeleine said, 'Miss Véronique, I don't think it's safe for a girl like you to be out on your own.'

'Because of the children going missing?'

'Partly that, yes.'

'But I am not a child.'

Madeleine stared at her with her habitual steady gaze. 'No one knows who's taking the children, or why. So no one knows who's safe and who isn't. And besides that, Paris has its dark side – full of people who, given half a chance, would rob you or trick you or worse.'

Véronique watched Madeleine as she continued brushing at the lace. She guessed Madeleine knew such people all too well; that they had been part of her world, part of the reason she couldn't go back.

'And there's a mood on the people,' Madeleine continued, 'the crowds won't be predictable. They're blaming the aristos today, but tomorrow it might be the bourgeoisie.'

'Then come with me,' she said quietly.

'Come with you where, Miss?'

'To the places girls like me aren't supposed to go. Show me this place of darkness.'

Madeleine looked up from her brushing. 'Why on earth d'you want to see?'

Because I'm tired of living in a cake box, and because I want to know what's really happening to the children. 'Because if I'm to create and recreate, I must understand humanity. I must understand life. I'd like to see the city as it really is.'

Madeleine was still frowning and Véronique felt foolish for asking. But then she looked at Véronique, her eyes clear, her face suddenly open. 'All right, then, Mademoiselle. If you're sure that's what you want, I'll show you the real Paris.'

They began the following morning, walking past coffeehouses, theatres and pâtisseries, to reach the Place de Grève, where a long line of men and women, some with shoes, most without, a few with babes strapped to their backs, stood waiting outside the city's Bureau of Poor Relief. Here too loitered gaunt-looking men, seeking a job, any job, hugging their arms around themselves against the bitter morning chill. The pillory stood empty, but it was here, Madeleine told her, that *macquerelles* would be stoned for corrupting young girls. It was here too that the executions were carried out – the gallows erected or the pyres built up – when there was a murderer to be hanged or a sodomite to be burnt or a traitor to be broken on the wheel.

On the Quai de la Tournelle she heard sellers preaching of miracle cures for pox, bad breath and misery. She saw men pulling teeth from people's mouths with pliers; she saw false teeth strung from cords. Madeleine took her to the fruit and herb markets of the Place Maubert and the stalls of the Rue de Bièvre, where cheesemongers in thin coats did fine business in goats' cheese, Gruyere, Parmesan and Brie. Just as with the previous day, Véronique could not shake the sense that some-one was watching her, but perhaps it was only that, closeted as she'd been, she was little used to crowds.

On the way home, they stopped at a chestnut-seller's who shovelled a pile of still-hot chestnuts into twists of brown paper. They stood warming their hands by the brazier, watching a man at a shabby desk on the roadside, writing on a scroll of paper while people waited in a row.

'What is he doing?'

'He's a public letter-writer,' Madeleine said, picking off the shell to reveal the creamy-white kernel within. 'Many people can't read or write, so they pay a few coins for him to write for them.'

Madeleine watched as a shrew-faced woman in a faded shawl jabbed her finger at the man.

'How do they know he's writing what they tell him to write?'

'They don't, I suppose. Or at least they won't until much later. He's unlikely to trick them, though, or he'd lose his business, and then he wouldn't survive.' A pause, and then Madeleine asked, 'Do you know what it is that your father's up to, closing himself up in that room?'

Véronique shook her head. 'No. He won't tell me. He says it's highly confidential. So you see, those men were probably right – I'm not to be his apprentice after all.' She tried to say it lightly, but the tears had risen in her throat.

Madeleine looked at her closely. 'Is it some experiment, d'you think?'

Véronique frowned. 'It could be, I suppose. All of his work is an experiment in a way. Why do you ask?'

The maid shrugged. 'No reason. I just wondered why it would be secret even from you.'

'Apparently I'm not "ready", whatever that might mean, and the King wants it kept confidential.'

'Maybe for the best, Miss. The King, they say, has very particular desires.' Her gaze shifted. 'Look! A marionette show.'

A makeshift puppet theatre had been set up on the bridge and people were gathering around to see two wooden puppets, crudely painted, dancing on their strings. One was dressed in black with a black hat and a white face, the other was smaller, its hair yellow straw, its limbs being jerked about by the invisible puppeteer so that it gave the impression of running.

'Help!' the smaller puppet appeared to shout. 'Help! The monster is coming for me!'

But it was too late for the marionette, as the puppet in black was upon him, 'Come with me, little boy! My mistress needs her medicine and there are no children at Versailles!'

'No!' shrieked the puppeteer in a cruel mockery of a child's screams, at that moment releasing red ink so that it spurted over the puppet and dribbled down its arms. And all at once

Véronique was back in chapel watching the blood flow from the statue's eyes, listening to the Sisters scream.

It was only after a moment that she noticed the tall man standing close to them in a dark brown cloak, and by then he was turning so that she couldn't see his face, but she had a strange feeling that she knew him.

That evening Véronique returned to her work on her doll, perfecting the movement of the eyes. She would not give up everything simply because her father had given up on her.

'You made this yourself?' Lefèvre was standing behind her, peering at the doll through his glasses. It was the first time she'd spoken to him properly in several days. 'It's very clever, my girl. Very lifelike.'

'You are kind, Monsieur, but it's only a toy.'

'A very fine toy. In any event, that's really all your father makes. But yes, Reinhart was right – you have talent.'

'But not enough to assist him on his project.'

His smile faded. 'That isn't the reason you are not included, my dear.'

'And what is the reason?'

'It is not, I'm afraid, for me to tell you.'

'But do you think it right, Monsieur, to keep me in the dark? For my father to leave off my education when it's only just begun?'

Lefèvre scratched his scalp beneath his wig. She noticed that the wig was singed, perhaps from his holding a candle too close while reading. 'I think you are still very young and that your father wants what's best for you.'

'Does he? I think he's forgotten I exist.'

And if he wanted what was best for her, why keep her in a convent all her life, ignorant of how the world worked, away from the vital life of the city? Why bring her here and begin to teach her only to shut her out again?

Lefèvre took out his silk handkerchief and used it to clean his glasses. He surveyed her closely. After a moment he said: 'The problem with your father, Véronique, is that he finds moving animals considerably easier to deal with than people, certainly vibrant young girls like yourself. And despite being lord of lifelike creatures, he has not mastered social artifice.' He smiled, appraising her. 'Perhaps I will give you lessons, Véronique, while your father is too busy to do so. I said before that I would teach you, and now Reinhart is not in a position to insist he will do so himself. Let me talk to him.'

'You would do that?'

Lefèvre smiled. 'I know what it is to feel undervalued, Véronique, to not have one's abilities recognised to the full. And I can see you are an open-minded sort of girl who will make an excellent pupil.'

'Thank you, Monsieur. I would like that very much.' She wanted to say something more, but her throat seemed to have closed over and she had a terrible fear she might cry.

'And I will ask Reinhart to talk to you, Véronique, to explain as much as he can. Perhaps I will talk to the King too. He may be reticent to have a woman involved, but I think in fact that you might be very useful.' He looked again at the doll. 'This would impress Louis, you know. Charm him. You could present it to him as a gift.'

Véronique pondered this. It had occurred to her before that she should seek out the King's favour. After all, if anyone could further her career as an automaton-maker it was Louis XV himself. Yet the man himself made her nervous, uncomfortable. 'But what would he want with a toy?' she asked. 'His daughters are too old for dolls now, aren't they?'

'Ah, but Madame de Pompadour has a young daughter.

Alexandrine. A skinny little thing who's recently been taken to convent.'

Véronique swallowed, remembering again her own arrival at the convent, the echoing of footsteps on stone. That decided it. 'Then I would like the doll to be a gift for her, Monsieur. I suspect she will need the company.'

12

Jeanne

A nick here, a tiny cut, just as the man had shown her. It was immensely satisfying to slice into something so precious, though Jeanne couldn't have said exactly why. The carnelian was gripped in a small vice so that it remained in place as she worked in her parlour, first sketching, then chiselling the design with the tiniest of drills. It wasn't yet perfect, but it would be. The engraver had shown her how to etch the picture onto the gemstone, working with ever finer instruments until the jewel was incised with the exact image: a carefully coiled snake.

As she worked, Jeanne went over and over in her mind the latest *poissonade* that – between the kitchens and her rooms – had made it onto her breakfast tray to ruin her morning chocolate. There were many who could have written or requested it, of course: the pompous Dauphin or his embittered sisters keen to remove her from their father's bed, the Duc de Richelieu, the brothers d'Argenson, or one of their many venomous followers. A 'trollop's bastard daughter' she had heard their circle call her; a leech, a vulgarian, a whore. She'd hoped that as time passed, the court would come to accept her as the King's mistress, despite her humble origins. She had exercised her powers of charm and flattery to the full. But one could not soften a stone. Instead, encouraged by the machinations of that shrivelled roué Richelieu and his clique, the courtiers had become nastier and ever crueller. Not to her face, of course – that would require courage, and men like them preferred to attack in the dark. No, it was far easier to try to undermine her by spreading rumours and malicious epigrams.

The latest was the most venomous yet: it claimed that, frightened of losing Louis' affections, she had taken to visiting a witch to buy potions to restore his love. Other verses she had tried to laugh off, but this was too dangerous. One did not joke about witchcraft at Versailles. The walls still held the screams of those tortured and tried at the *chambre ardente*, accused of poisoning, malefecium and murder. People still whispered of how Athénaïs de Montespan had used the dark arts to win the Sun King's love: blood-soaked ceremonies, infant sacrifice, all claims that were never proved. This might be the age of supposed rationality – of sweeping away the cobwebs of superstition and darkness – but they were still there lurking in the corners, waiting to creep back out.

Jeanne laid down her tools and went into her dressing room to rinse her hands and neaten her hair, noting that, even without carmine, her lips were pleasingly red. Miette, her monkey, was sitting at the dressing table eating a plate of figs. Jeanne took the one that she now held out to her and stroked her little head. 'Let us hope, *ma chérie*, that Louis hasn't heard about that particular *poissonade*.'

He would be here soon, annoyed after the *Lever* where each morning he must pretend to rise from his bed before the entire court and be subjected to the ridiculous ritual of washing and dressing and shaving and prayers, the First Valet passing the royal mirror to the Grand Master of the Wardrobe, who passed it to the Dauphin, who passed it to the King. From his waking to his retiring, attendants hovered around Louis like flies about a corpse; his every mouthful, comment, every bodily function a matter of intense debate; everything he said, drank, ate or shat compared to the excrescences of the Sun King. 'Solid this morning,' she'd heard a surgeon observe yesterday. 'Surely an auspicious sign.'

Here, however, in Jeanne's apartments, in her company, he believed himself to be free. She'd considered it one of her greatest achievements as the *maîtresse-en-titre*, one that made her indispensable to his needs. But she saw now that she'd been wrong. As Jeanne wiped away some of the kohl from beneath her eyes, she knew that despite the potions and powders she applied each day, despite her continual efforts to be lively and witty and entertaining, Louis was losing interest in her, just as he'd tired of that poor bitch the Queen some years ago. And her enemies, scenting blood, were moving in like a pack of wolves.

What should I do now, then, Mother? she asked the jewelled mirror. *What should my next move be?*

For as far back as she could remember, Jeanne's mother had called her 'Reinette', the girl who would become the King's mistress. It had been predicted by the fortune teller, and so it must come true. She'd trained her for years in the arts of conversation and flattery and manipulation until her role at court had seemed almost inevitable, despite her bourgeois beginnings and her absent father. 'The most important thing, *ma chérie*, is belief in yourself, in your own value. Most people will take you at your own estimation, so make sure that you set the price high.'

But now her mother was dead and cold and there was no one to tell her what to do. If Jeanne could not keep the King in her bed, she would lose her position, no matter how well she kept him entertained. A mistress's first duty was to fuck. And if she were dispensed with, what then would she be? Nothing. She'd left her husband years ago. She'd had no choice, her mother had told her – it was her destiny, her role, her place. Charles, however, had been rather less understanding. The male ego was a fragile thing, more delicate than her Sèvres porcelain. He wouldn't accept her back, wouldn't even acknowledge her. She would be cast out of Versailles, out of Paris, just as poor Louise de Mailly had been, left to mourn to death in her shabby clothes while Louis tupped her sisters.

But no, she was being self-indulgent; she was not meek little Louise. She would not be dislodged so easily, and as luck would have it, she had no sisters; only a very handsome brother, and that wasn't to Louis' taste. She just needed a better strategy; some further strings to her bow, some poison for her arrow-tips.

'There will always a way to keep your power,' her mother had told her before she died. 'Another way to ensnare him. The good thing about this particular Louis is that, though he has great pretensions to learning and science, he isn't terribly clever.'

The door opened. She turned.

'Look at this, *ma petite*!' In his hand, Louis held a doll, which he set down on Jeanne's dressing table. It was a rigid figure upon a base, holding a book in its hand. Jeanne stared at it, then raised her eyes to Louis in question.

'It was made by the clockmaker's daughter. For you to give to Alexandrine.'

'How ... thoughtful of her. I'll take it to her tomorrow, when I go to the Assumption.'

'She's happy there?'

Jeanne blinked. Sometimes the man's wilful blindness astounded her. 'Hopefully she will be in time.' It was only a month since Louis had decided that Alexandrine, now six, should be educated in a convent. The nuns reported that she still cried every day. But this, of course, was just the way things were at Versailles: no laughter of children playing, no cries of babes. The only children at Versailles were cast from bronze or carved from marble. Not even Louis' own daughters had been allowed to stay here as children, none save for Adélaïde, who had begged and wept until her father gave in. Jeanne would never beg and weep. It would only incur Louis' wrath. She'd seen his anger laid bare before and she'd no wish to experience it again.

She stared at the doll. It was a curious-looking thing with a face carefully painted to look like that of a real girl. She flinched. Had she imagined it, or had its glass eyes moved?

Louis laughed. 'Clever, isn't it? You press here and the eyes move from side to side, as though she's reading her little book.'

Jeanne watched more closely as the glass eyes slid about in the porcelain face. The doll was uncanny, horrible. She would never give this to her daughter. 'How delightful,' she said. 'How ingenious.'

She wondered if Louis recalled the last time he'd given a doll as a gift and how unfortunately it had all ended: the little Infanta Mariana Victoria, presented with a doll bedecked with precious stones by the cousin she was one day to marry. Only then they decided they needed an heir swiftly and so the Infanta was bundled off back to Spain and the already fecund Marie Leszczynska was drafted in to commence her many years of breeding.

'You should see the things Reinhart himself makes. This is but a toy compared to his creations.'

'And his daughter is learning his trade, is she?'

'He is training her up for a specific role, I understand.'

'That's ... original.'

'She *is* quite original, I think. She has returned only recently to live with her father, who's been teaching her, but according to Lefèvre she is already capable of things I myself am not. He says she is exceedingly intelligent for her sex.'

'I see. How old is she, this girl?' She already knew, of course. She had read it in the police reports, seen the girl's perfect skin with her own eyes. But she wondered if Louis was aware.

'Seventeen, I believe, though you'd think her older. There's little of the silly girlishness you see in most of that age.'

Jeanne twisted the cup in her hands. She did not like the tone of his voice, nor the set of his lips. If Louis were to have other little birds, then far better if they were birds she herself netted: pliant, docile girls who could be bought off and stowed away. Not clever clockmaker's daughters who might covet her own role.

'Well, one can never really tell with girls of that age how they will turn out, can you?'

'No, indeed,' he said, his tone suddenly gelid. 'At seventeen you would have been merely a bourgeois upstart. And look at you now, so very high up.'

She smiled, but beneath the paint her cheeks flamed crimson. She knew exactly what Louis meant: so high up she teetered on the edge. So high up she could fall.

'I hear there is another *poissonade*,' he said. 'You did not tell me.'

A shard of fear. Who had told him? Du Hausset? Surely not. 'I did not like to trouble you with such nonsense.'

'And I do not like to have things withheld from me,' Louis said, and she knew his meaning was double. He took her chin in his hand and held it firm. 'I do not like to feel I've been misled. Do not make me doubt you.'

After Louis had left for Mass, Jeanne drank a cup of wine to steady herself and had du Hausset lace her tightly into her whalebone corset, as though it might keep her standing up-right. She needed to recover her wits, for she had few other weapons. Wearing only her short petticoat and mauve satin powdering gown, she seated herself at her dressing table, as it was nearly nine o'clock, the hour of her first formal toilette. It was Wednesday, the day Lieutenant Berryer came to update her on police business. After all, it was she who had secured his appointment; it was only right that he reported first to her. And she needed to talk to him today about the *poissonades*. She needed to form a plan.

As Jeanne's hairdresser, Bertin, set the curling irons to heat the fire, Berryer entered the room, bowed and placed his port-manteau on the floor.

'Marquise.' He kissed her hand. 'You grow younger by the day, I swear on it. Your eyes ever more sparkling.'

Jeanne smiled at him. In the three years since he'd assumed the role of Lieutenant General, Berryer had aged at least a decade, his proud face now etched with dark lines. It was a role with many responsibilities, of course, and many secrets, but it was also a role of great value: it was what he had wanted. She would not blame herself for the white hairs in his brows or the bruised pouches beneath his eyes.

'Tell me, then. What news? Have you managed to find evidence against the author of those horrid little verses? You got my note saying another had arrived this morning?'

'Yes. A horrible slice of malice. We do not have evidence as yet, Marquise, but I'm confident we will have it soon. The *cabinet noir* are about their business of opening and resealing letters and I believe that within a few weeks we'll have tracked the author or authors of the vicious *poissonades*. There aren't many who have the motivation.'

'On the contrary, Nicolas. There are many who wish me ill, as you well know.'

Berryer turned over a sheet of paper. 'They wish you ill only because they know your influence continues to grow.'

'And with it yours.'

He bowed again. 'Marquise.'

She smiled, yet she was sure Berryer knew her position, for his spies burrowed everywhere. He knew just as she did how Louis' eye roved to the many pretty young women at court and beyond, none of whom would refuse him. He must know too of how she couldn't continue to satisfy Louis' appetites despite the rest Quesnay had insisted she take, despite the diet he'd prescribed: vanilla, truffles, chocolate, celery. They merely made her sick. But Berryer knew something else about her too: that she was just as ambitious as he was – more – and that to leave Versailles would kill her. She had spent her lifetime climbing to this point. If she fell now, she would die.

'What developments in Paris?' she asked.

'We continue to be besieged by paupers and low-value persons, driven from the countryside by the cold and the poor harvest and seeming to think Paris has fires forever lit. However, my officers have had some success in dispersing them in accordance with the King's edict.'

'Yes.' Jeanne kept her face smooth but felt her heart dip. A war on vagrancy, they called it, calling a halt to moral decay. But did one go to war with one's own people, simply because they were poor?

Bertin brought forward the curling irons and began to take pieces of her brown hair between his fingers and smooth them.

'Have there been any difficulties?' she asked Berryer. 'Has the King's edict proved popular or turned people against him?'

'I suspect, Marquise, that few know or understand about what was decreed back in November. However, they must know just as well as we do that the capital can't continue to absorb the rural poor. The hospital and poorhouses are already besieged.'

'Then the people still love him?'

'Of course, of course.'

She did not believe Berryer. Louis' status as *le Bien-Aimé* had been badly dented, by the costly war, the unfavourable peace. She'd read in the police reports of how the people blamed her, as much as they did Louis: a parasite on the wealth of the state, poison in the ear of the monarch. She knew too how d'Argenson insinuated to Louis that she herself should be sacrificed as an offering to a hostile France.

'Are there any other matters I should be aware of, Nicolas? Any unrest, complaints?'

He squirmed.

'It is better that I know.'

'A foolish matter, truly. The only cause of disquiet, aside from the usual grumbles, is a few complaints about pauper children being taken.'

A twist in her stomach. 'Taken where?'

'It isn't known. There've been a clutch of claims about children in Paris simply disappearing, but I suspect this is the stuff of rumour, the popular tongue exaggerating one or two instances of children who have, in all probability, run away or perished. I am not concerned.'

'Well, perhaps you should be concerned. It's the sort of thing that will lead to discontent; to questions being asked. Who do they suspect?'

'There are a variety of theories, it seems. Some say it is migrants from the country leading children into corruption, but the most popular is some drivel about an aristocrat snatching children from the streets and draining them of blood. And they say this is the age of reason.'

'I do not like this. It needs to be dealt with.'

'My men are already looking into it, Marquise. But as I say, I believe this is a case of rumour overtaking reality – of the Third Estate developing reports of a few runaways into a deluded conspiracy about the nobility.'

'But that is in itself dangerous.' She put her hand to her throat. 'It will lead to social unrest.'

'As I say, it will be smoothed over. You need not worry yourself, Marquise.'

Her eyes alighted on the horrible doll, lying on the corner of the table. As if she would take such a thing to Alexandrine. 'Your little *mouche* who's keeping an eye on the clockmaker – what has she said about the man's daughter?'

Berryer, taken unawares, gave a slight frown. 'I don't think she's said much at all about the daughter. An innocent little thing so far as I'm aware.'

Innocent. Louis would like innocent. They were less threatening that way. The doll, though. The doll was strange. Maybe not as innocent as all that.

'Do you think her pretty? She's slightly unusual-looking, wouldn't you say?' She watched as Berryer's complexion went from white to red. Yes, he thought her very pretty then.

'Only in the way that all young girls are pretty, Marquise.' He recognised his mistake as soon as he'd made it. 'Which is not to say—'

'Nicolas, I know this too well.' Jeanne picked up a golden pot of rouge and inserted a finger, drawing a small circle high on her right cheek. She no longer seemed to have her own colour, as though life at court had drained her to grey, as though all the blood had been leeched from her veins.

'What else? Your men have caught no hints of any movement against Louis? Nothing that threatens his safety?'

'Nothing. Rest assured I would tell you immediately.'

Jeanne watched in the mirror as Bertin folded a curlpaper into place and lifted the hot irons. As he twisted one of her light brown locks there was a huff of smoke and the faint scent of singeing hair. 'I hope so. I hope you would tell me every-thing, no matter what it was.'

Berryer nodded slowly. 'Of course, Marquise. Of course. It's always you I come to first. I remain, as ever, your servant.'

When Berryer had gone, Bertin removed the curling papers and began separating the new-made curls with his fingers.

'You are still the most beautiful, Marquise. There is no one at court who can touch you. That is what they all say.'

She smiled thinly. Oh yes, she knew what they would be saying now: that she kept herself beautiful with black magic, just as Athénaïs de Montespan was rumoured to have done the century before. That she preserved her youth with the blood of virgins, the ground-up bones of babies. 'Thank you, Bertin. The golden bees today, I think. Something to give me a sting.'

As he sought out the jewelled pins, she picked up the reading doll. The book which it held in its hands was, she saw, a miniature version of Perrault's Cendrillon, the girl who had risen from the kitchens to become a princess. Was that a comment on her beginnings? A suggestion she could be swept aside? It was too clever, too strange. She set it down, no longer wanting to touch it.

'Bertin, when you've finished, get rid of this, please. Discreetly. I don't want it in my rooms.'

'Yes, Madame. And now, should I admit the others?'

Jeanne closed her eyes for a moment. She could hear them already – ambassadors, admirers, bidders for favours – assembling on the stairs for her second toilette, where she would go through with the ritual of applying a second layer of face paint, a second set of pins, and listening to their flattery and lies and requests for approval. She ran her hands through her hair to give the curls a more natural appearance and wiped off some of the rouge.

'Yes, Bertin. I suppose it's time. Let them all come in.'

13

Madeleine

The bastard bells of l'Église de l'Oratoire tolled again, and again – Sunday morning. Nine days had passed since Reinhart had been appointed *Horloger*. Yet apart from the whispered message from the man in red, Camille had sent no word: no instructions, no sign, no blunt, and with every day that passed, Madeleine grew more angry, more jittery, until she could have wrung the man's scrawny neck. Thirty days, he'd said. A month at the clockmaker's and then she would be free – the money would be hers! And she'd been so nearly there, nearly out, only to be told by some sullen-faced guard that no, she was not allowed to go at all. The state had played with her like a cat played with a mouse, offering freedom, but then clawing her back the moment she tried to run. Though the thought turned her stomach, she knew she had to return to the Rue Thévenot – she had to find out what Maman knew.

A fog had set in over Paris, made up of last night's rain, last night's fires, the fumes of the stinking sewers. It rose in drifts from the gutters and seeped along the alleyways, obscuring Madeleine's path as she walked along the Rue Plâtrière towards the place she'd once called home. It wasn't just Camille that was worrying her, it was the clockmaker himself. Ever since the King had visited, Doctor Reinhart had remained closeted in his workroom, muttering with Lefèvre and scribbling plans and letters, not even admitting Véronique. When they were out, the workshop door was kept constantly locked, meaning she knew she had to get in. Madeleine had watched, silent as a shadow, as Reinhart turned the key and slipped it into his waistcoat pocket, but it was Joseph who looked after his master's clothing and so far she'd failed to get hold of it. Madeleine's only opportunity to get into the room was when she was rung for (to bring them coffee, brandy, almond biscuits) and then the men fell silent, or turned abruptly to another subject. Books would be closed, papers shut away. The snatches of conversation she heard were mainly of dreary technical things: obtaining some substance from trees to use as tubes, speaking to some physician about obtaining original samples, getting plans some other automaton-maker had made. The sketches Madeleine glimpsed then meant little to her – mere lines and numbers, seeming to show the inner workings of some machine. Some of the things she saw, though, made her wonder: a set of bellows, a long syringe, a model of a human hand. She saw nothing of the invention itself, but through the walls she could hear the whir of the lathe and the scratch of Reinhart's saw. Whatever they were making, they were hiding it.

She had reached the Rue Thévenot now. Six weeks it'd been since she first left home and in that time the street seemed to have narrowed, darkened, shadows pooling in the doorways and in the hollows of people's cheeks, the ammonia stink of piss bleeding from the buildings and gutters. Maman's house too seemed shabbier, poorer, with its cracked facade and worn front steps, the shadowy pathway to the back. She didn't want Émile living here. It wasn't safe, certainly not now that there might be a child thief lurking in the shadows, waiting for the unprotected poor.

'They're not up yet,' the little servant who opened the door told her. She was a straggly orphan girl Maman had taken from the Hôtel Dieu, barely eleven years old. She'd bad teeth but fair hair and a pleasing enough face. A *beurre demi-sel*, as the punters would have it: a girl halfway to a prostitute.

The fires were only just lit, and poorly. Madeleine used the poker to turn the embers. 'Go wake Madame now, would you? Tell her Madeleine's here.'

When the girl looked nervous, she pointed the poker at her and said kindly, 'Tell her I made you. That I'm terribly mad. That I threatened you with this stick.'

Maman appeared shortly afterwards, her unmade-up face like a ball of dough left too long on the shelf. 'This'd better be good, Madou. I need my beauty sleep.'

'And I need to talk to you. Coraline too. It was the only time I could come.'

Maman commanded the maid to bring them all chocolate and took Madeleine through to Coraline's room, where the air was thick with her sister's scent, and that of the man who'd been here: hair oil, semen, sweat. Maman dragged back the curtains to let in the greyish light. 'Up, girl! Your sister's here.'

Seeing Coraline half asleep, her face smooth, Madeleine could remember what she'd been like as a child, before Maman had started her 'Academie' of vice, before Coraline had grown scheming and cold. What would she have been like if they still lived above the pet shop? If she'd married some clerk and moved out of town? Would she have been a warmer elder sister, one who comforted Madeleine when she cried rather than telling her to toughen up? Perhaps she'd have always been cruel and calculating. After all, she was her mother's child, and her mother had any tender feelings cauterised in the womb.

Coraline's face broke into a yawn. 'What time is it?'

'Time to tell me what's been going on. I've been left, you see, in the dark.'

'About what?'

'About what the bloody hell the police are up to. About why, nine days after I finished the work they set for me, I'm still not allowed to leave. Or had you not noticed that I'd not returned? Had the time simply passed you by?'

'We knew you was at the Louvre,' her mother said, 'because you wrote, didn't you, saying you'd done it? But you told us precious little else. So what happened, then? Tell us everything. Tell us what you found.'

Madeleine told them then of the former kitchen girl who'd claimed Reinhart had drawn her blood. She told them too of Reinhart's appointment, Pompadour's visits, the whispers of the man in red.

'Madame de Pompadour,' Coraline breathed, moving over to her dressing table. '*Bon sang*. To think you actually met the woman. Tell us exactly what she was wearing. How did she have her hair?' She piled her own curls up onto her head as she stared into the stained looking-glass.

Madeleine, standing against the wall, scowled at the back of her sister's head. 'Forgive me for not covering the fashions, but I'm rather more worried about where my bloody money is and what the police have in store for me. I'm rather more worried about how a kitchen girl who made claims against him wound up dead.'

'What do the other servants say about her?'

'I've got little out of them. They only say they thought the girl was a thief and a liar.'

'Well, then. What are you worrying about? A girl'll say anything to escape the noose.'

'All the same, Camille kept it from us, didn't he? That rather vital piece of information.'

Maman looked away and it came to Madeleine in a rush that of course her mother had known, had known exactly what risk her daughter was taking, and had sent her into battle, blinkered.

'*It'll work out just fine, you'll see,*' she'd said to Madeleine, the day before she'd left for the clockmaker's. '*Make sure you keep your wits about you, and no harm will come to you, I'm sure of it.*'

Only Maman had told her that before, and Madeleine had been harmed. She should've known not to believe her.

'You've not seen Camille, then?' she asked her sister. 'When was he last here?'

'Haven't clapped eyes on him for over a fortnight. Can't say I've been missing him terribly. I see why Suzette didn't like the man – he's one of them likes to leave his mark.'

'Thank you, Coraline,' Mother said quickly. 'Madou didn't come here for that.' She sat back on the divan and rearranged her bosom. 'Most unsatisfactory, I agree. I never like a man who doesn't pay on time, nor a man who draws it out too long. Could of course be that he's just busy, or lying low for a while, but I don't like it.' She inspected her cup of chocolate and grimaced. 'We'll find him out, Madou, you be sure of it. We'll make him pay.'

'How?' It struck Madeleine now that it'd been foolish to agree to anything before seeing the money. Why had Maman, ever the businesswoman, not seen that? 'If Camille decides he won't pay me, what are we to do? Go to the police? Go to Berryer?' She laughed.

Maman raised her pale eyes to hers. 'You forget yourself, girl. You forget who you speak to. You think I don't have other friends I can call on, other favours I can call in? Camille will pay all right. And then we can cut the shine.'

Madeleine responded with the thinnest of smiles; Maman

had already had her cut. It was carved through her very face. She should never have come back to her mother's house. She should have found Camille by herself.

'Now, don't you give me that look,' Maman continued. 'All will be well, and God willing you'll be home safe in a week or so.'

'Yes. God willing,' Madeleine replied coolly. As if God had ever taken any interest in what happened in this house, or in any house like it.

'We'll get your room ready for you, eh? Put some new prints on the walls. Make it nice.'

Madeleine forced a smile, but she had no intention of returning to this seedy house with its yellowed wallpaper and greasy beds, its grimy rag-plugged windows. The air was filled with memories she'd tried to bury, and the idea of having to return here made her sick. It was more than that, though. Since escaping Maman's, since meeting Véronique and Doctor Reinhart and Joseph, something in her had changed, grown, like roots burrowing beneath the soil after lying long dormant. She couldn't return to what she'd been. And she couldn't let Émile remain there, to grow dead inside just as she had.

Walking back into the smoke-stained hallway, Madeleine leant close to the servant girl, smelling her scent of coal dust and neglect. 'Don't stay here, eh, *chérie*? Get another position. Better most other houses than this one.'

She found Émile in the street, playing quoits with two other boys, scrawny, snotty little things with strips of leather tied about their feet. He was sullen and quiet when she tried to talk to him so that she worried at first that he was ill again, or hurt, until he blurted out:

'Thirty days, you said! Thirty days and you'd be back. It's been much longer than that. I've been counting.'

Madeleine's heart squeezed, the old feeling of guilt spreading through her. 'I know, Émile. I thought it *would* be that long. That's what I was told. But I was tricked – I'm not allowed to leave yet.'

'*Why* not?' His pointed chin jutted out but she could see the tears glistening in his eyes.

'I don't know, *mon petit*. That's what I'm trying to find out, what I've been asking Maman. But it won't be much longer; it can't be.' Lying. Again she was lying. She'd no idea what was to come.

'It's because you want to be with that pretty lady, isn't it? You like her more than me now.'

'Miss Véronique, you mean? Don't be silly!'

But it was true she'd grown fond of her, despite herself. She had to hand it to the girl, she'd not flinched from any of the things she'd shown her this past week – not the *culs-de-jatte* begging with their stumps of limbs, nor the coffins stacked in the Cimetière des Innocents, the newly dead on the bones of the old. Of course, she hadn't shown her the very darkest parts of Paris: the bathing houses and seraglios which catered for every vice, the rookeries where families lived squeezed together in squalor, disease and damp, the worst kind of bawdy-houses where the dregs of society stewed together in misery. Nevertheless, Véronique was of much sterner stuff that Madeleine had at first reckoned, and with a clarity of ambition she wished she had for herself.

'I've grown a little fond of her, yes, but she's not kin, is she? She's not you.'

'I'm not staying here without you, you know!'

A shard of ice pierced her. 'What d'you mean, Émile?'

'I mean I'll go somewhere else, with the other boys. I hate it here on my own.'

She grabbed his shoulders and brought her face very close to his. 'Émile, you mustn't say that. You mustn't go anywhere. And you must stay near the house, like I told you.'

He tried to shake free of her.

'Émile! There is someone in Paris taking children. You know that?'

He stopped struggling and stared at her, the tears spilling from his eyes. 'Is it true, then, what they say about a monster chasing boys through alleyways?'

'Who says that, Émile?'

'Some of the other children. They say there's a monster who drinks blood. That he comes in the night in a black coach with a black horse and chases boys through the streets.'

Fear brushed against her like the thin wings of a moth. 'There's no monster, Émile, all right? No monster. That's just stories. But there might be nasty people about. So please no more talk of running away?'

'You don't know any better than I do, Madou. You're just as scared as me!' Then came another fit of coughing.

Madeleine took out her handkerchief and held it to his lips. 'I'll come and get you as soon as I possibly can, I promise you. And I will not leave you.' She wrapped her arms around him, around his thin sparrow shoulders. 'I am trying to get us out of here, Émile, like I promised your mother I would. You just have to stay strong while I'm away. I'll come back here as soon as I can.'

Stay strong. How many times had that been said to her? And what had it turned her into?

★

The fog was clearing as she walked back to the Louvre, exposing the smoke-stained buildings and unswept streets. High in the sky a white sun hung like a pearl. On a wall she saw a poster, freshly pasted over the mildewed mush of notices beneath. 'Le Foire Saint Germain' was printed in bold black type. How could festival season have crept up on her unawares, she who'd always looked forward to it; always taken Émile and Suzette to dance and eat and laugh? Perhaps because it didn't feel like a time to be celebrating, but rather as though a disease was flowing through Paris, a pustule ready to burst. All the same, she'd tell Véronique as soon as she got back, for it'd be something she'd want to see: the masks, the dancing, the food, the fights, the drunken and desperate revellers.

When she reached the apartments, however, she found Véronique was far too excited to listen, wanting only to speak to her of something else entirely.

'I've so much to tell you, Madeleine. Monsieur Lefèvre is to give me lessons, in anatomy and such. The King himself has insisted upon it. My father has agreed to it. Can you believe it? The man who tutors the King is now taking me on as his student!'

'Certainly it's a grand thing,' Madeleine said unsurely. From what she'd seen of anatomy lessons, she could imagine little worse. She'd had her fill of men's naked bodies, be they alive or dead.

'And there's more.' Véronique seized Madeleine's hands, her eyes gleaming, her cheeks flushed. She had the manic sort of energy that Madeleine had seen now and again in Reinhart. 'Father has agreed that I should help with their project. The King *himself* has requested it. He thinks me talented, Madeleine. That I have great potential.'

'Does he, indeed?' Potential for what, exactly, Madeleine wondered. She'd seen the way the King looked at her mistress and it was a look that she knew well.

'I think this is going to change everything, Madeleine.'

And perhaps it would. 'Do you know what it is now, then?

What they're making?' she said lightly. 'What is it you're help-ing them with?'

Véronique's smile vanished and she looked away. 'Well, that I can't tell you. It's all to be kept quite dark, but it is something significant, Madeleine, something that takes my father's work to a new level, and something in which I have a special role.'

Madeleine kept her silence, but she didn't like it. Why did they need Véronique to help?

Her expression must have betrayed her, as Véronique smiled. 'Oh Madeleine, do try to be pleased for me! Please try to understand. If I can be valuable to them, if I learn more than any of those wretched boys, if I can impress the King himself, then my father will keep me on as his apprentice. You see? He'll have no reason to return me to the convent. My future begins to look more hopeful.'

There was a nervousness in her voice that made Madeleine herself fearful. Véronique must, like her, wonder why Reinhart had changed course so abruptly, from shutting his daughter out in the cold, to bringing her in on the secret. She must, like Madeleine, suspect the King's interest in her had little to do with her abilities.

'What was it you wanted to tell me, Madeleine? When you first came into the room?'

'Oh, only that there's a fair starting soon. In the Rue de Tournon. It's always a colourful affair. I thought you might like to go.'

'Well, yes, I would, but I'm afraid we won't be able to con-tinue our escapades for a while. My father needs me to start at once. In fact, I am to go to Lefèvre's now.'

'Of course.' Madeleine felt the familiar blow of disappoint-ment and, within it, a pinpoint of fear.

She walked slowly back downstairs, sliding her hand along the smooth banister. At the bottom of the steps she stopped outside the workroom door. She didn't know what their project was, but she knew in her guts, her lungs, her blood, that Véronique should have no part in it.

14

Madeleine

Wednesday. Washday. Madeleine took some of Véronique's shifts into one of the Louvre courtyards to dry. Another maid was already stretching cotton sheets over the rosemary bushes and nodded to her in greeting. As she worked, Madeleine considered the tree growing close to the wall. Could she somehow climb it and enter the workroom though the upstairs window? Three days had passed and still the door remained locked, still she remained in the dark. She'd seen little of her mistress, for she was forever at her lessons with Lefèvre or helping the men with their curious machine. When she returned she seemed tired and disinclined to talk. Madeleine felt shut out, the door closed against her. She'd been relegated again to a servant.

'Victor has gone.'

Madeleine turned to see Joseph, his skin slick with sweat. She removed a peg from her mouth, her heart falling. 'What d'you mean "gone"?'

'I went to find him in the Rue Sainte Catherine and they told me he has not been seen for two days – that he's run away.' Joseph shook his head, still breathing heavily. 'That boy has not run away. He's been taken, just like the other children were.'

Madeleine's felt her chest constrict. 'Have they told the Watch?'

'Oh yes, because they think he's fled. But what do the Watch do? They do nothing about the other children. This is what people are saying.'

'He could have run away though, couldn't he? That bastard of a master of his—'

'Trust me. He did not run away. He knows what they do to runaway slaves.'

'What do they do, then?'

'Branding, beating, whipping. If a person is not considered truly human then you can do anything to them.'

For an instant Madeleine remembered the footsteps behind her, being pulled to the ground as she tried to open the bedroom door, her fingers scrabbling at the wood. And then the smell of burning metal.

She put down the basket of washing. 'We'll go together, now, and search. Someone must've seen him, mustn't they?'

'Perhaps.' Joseph took out his handkerchief to wipe his face. She had a sudden urge to do it for him. 'Or perhaps it's already too late. He could be on board a ship. He could be anywhere.'

They hurried to the Marais district, past the elegant houses and high-walled *Hôtel particuliers* of the opulent Place Royale, to places Joseph thought Victor might have visited: the paperworks, the market, the Église Saint Paul, houses where servant friends worked. Madeleine had never walked out in the streets with Joseph, never seen before how others reacted to him, as though he were a different species, one at which they could gawp or laugh or spit. Oh, she'd had her fair share of glances in her time for her damaged face, her fair amount of jeers for just for being a woman, walking alone in Paris, but this was something new. And Joseph bore it all as though he was walking about in a full suit of armour. She wanted to say something, but she didn't have the words. She only walked a little closer to him and scowled at anyone who stared.

For two hours they trudged the streets, but no matter where they went, no one had any real information to give them about what'd happened to Victor. They heard only rumours and whispers of other vanished children and theories as to where those children had gone. A blood-stained butcher at a stinking shop in the Marais told them, loudly and with the confidence of a man holding a cleaver in his hand, 'This isn't just pauper children anymore, is it? It's errand boys and apprentices. Working men's children. And the working man, and working woman, won't be having that.'

He was right, Madeleine thought. Berryer had sneered at her questions and talked of runaways, but he couldn't shrug off the rising demands of a crowd of angry traders.

'If you ask me,' the butcher continued, 'this man in black clothing people've been talking about? He's police. A police spy.'

Madeleine's eyes were on the man's hands, the knife, the black blood beneath his nails. She was aware, though, that a crowd was beginning to form in the narrow street around the butcher's, of tradesmen, shoppers, passers-by, a hopeful-looking dog.

'And that's why the Watch do nothing,' a woman agreed. 'Because it's the Watch themselves that are helping them!'

Nods and jeers from people standing nearby, and the butcher, warming to his subject, continued, 'And who would be instructing the police to go picking up little children? Why is no one speaking out about what they're being asked to do?'

'Because it's the nobles that are paying them for the children!'

'It's a leprous prince,' another woman interjected. 'That's

what I heard. A leprous prince who must bathe in blood to get cured.'

The butcher nodded, leaning forward. 'I wouldn't be surprised. The rich have been draining the poor for years. Now they take our children – our own flesh and blood.'

Madeleine, now deeply uneasy, elbowed Joseph in the side. 'Let's leave this place,' she muttered. 'This lot don't know anything about Victor, and they'll only attract the Watch.'

'You go back,' Joseph replied. 'Tell Edme what has happened. There are other places I can try. I mustn't give up yet.'

Madeleine didn't notice the man outside the Louvre until she was within a few paces of him. Her heart gave a sickening lurch as she realised that it was Camille, leaning against the wall, sucking on a long clay pipe.

'She lives, and yet she does not write to me.'

Madeleine felt her stomach turn to acid. 'I've been waiting to hear from you,' she whispered.

Camille gave a look of mock surprise and moved closer to her. 'Indeed? I believe I was quite clear. I said you were to report to me at least once every week. Yet over two weeks have passed and' – he held his hand out, palm upwards – 'nothing.'

She tried to steady her breathing. 'Reinhart was made *Horloger*. The job you asked me to do is done. I'm owed my money.'

He smiled, removed the pipe from his mouth and, without warning, grabbed Madeleine's arm and pressed the pipe's glowing end to the soft underside of her wrist.

She gasped and pulled away from him, clutching at her arm, her nostrils full of the smell of burning flesh.

Camille's voice was low and hoarse. 'You work until I tell you to stop. Until I stop winding the key.'

She gritted her teeth. 'I did as you asked. I got the information you needed and they appointed Doctor Reinhart *Horloger*. I'm owed my money and I need to get back to my nephew. I promised him I'd only be away a month.'

'You'll be paid when you've proved your worth.'

'You said—'

'I *say* that you'll continue to report to me and that you'll do precisely as I tell you. If you don't?' He shrugged. 'Well, you know what happens to police spies who are deemed to have refused orders, don't you? They go to the Bicêtre. Mostly, they don't come out.'

Madeleine stepped back. The Bicêtre was the worst prison in Paris – it was where they sent the murderers, the diseased, the violently insane. 'You wouldn't.'

He shrugged. 'I would. If I had to. And then what would happen to that little brat of Suzette's? Who would he run to if you were locked up for an indefinite length of time?' He kept his eyes on her, cold as a reptile's. 'Or you could just do as I ask.'

Madeleine felt anger rip through her then, like a flame. He'd tricked her, of course he had, just as he'd tricked Agathe. Five hundred livres, held out like a bag of sweets. And she'd been so eager to escape, so eager to save Émile, that she'd gone along with it, believed it. Now he'd use Émile against her, as a bargaining chip, a way of making sure she complied. Tears of fury welled in her eyes and she cursed herself. Why must rage make her cry, rather than shout? Camille would think he'd won.

Sure enough, he took out a flask, poured water on his handkerchief and began to wrap it around Madeleine's wrist. His voice softened. 'It's not so much to ask, is it, now? While many in Paris starve, you live here, comfy as a cat. You've regular employment, a kind mistress. All I ask in return is that you keep me informed of anything important, and whatever Reinhart is doing now. It matters.'

'How d'you know that?' she said stiffly.

'Because I know who his orders came from.' His hand was still on her wrist.

'Then why not go to that person straight off?'

'Ah, because that's not where *my* orders come from.'

It was still Pompadour who instructed him, then, and even she didn't know what Louis had asked Reinhart to do. That was odd, no doubt about it.

Seeing that she understood, Camille leant closer so that she could smell the tobacco and stale wine on his breath. 'He has some strange predilections, our King. Odd tastes. Can't necessarily be trusted to know what's best for himself. Speaking of which, how's your mistress?'

Madeleine tried to pull back. 'Perfectly well.'

'Is she? Are you so very sure?'

'Why d'you want to know?'

'It doesn't matter why I want to know, it's your job to tell me.'

'Tell you what?'

'Where she goes, what she does, who she sees, what she says.'

'She goes only to her lessons, to see her father and Monsieur Lefèvre.'

'You're quite certain about that?' He raised his eyebrows. 'I'm not sure that you are, you know.'

Madeleine thought back to Véronique's behaviour over the past few days – the silences, the absence.

'Watch her more closely. Follow her. Read all her letters. Go through her clothing. Never let her out of your sight and write absolutely everything down.'

She didn't like this. Why the sudden interest in Véronique? 'Why? Who wants to know?'

'It doesn't matter why or who. You are simply to do it, and within the week.'

'I'll have nothing to report, Monsieur. My mistress is a green thing, like you said.'

'And grown fond of you, hasn't she? How do you think she'd feel if I told her what you really were?'

Madeleine felt her blood chill.

'Yes. You think she'll be your little friend then, do you? And what do you reckon her father would have to say if he'd found out you'd tricked his household? What do you think the other servants here would do if they found a parasite in the Louvre? You think the people of Paris would act with restraint on finding a police *mouche* in their midst?'

Madeleine thought of the flood of people who'd arrived that morning when the butcher started shouting; the claims that the police were to blame. It didn't matter why she'd taken the job or acted as she had; the way things were going in Paris right now, she wouldn't be given a chance to explain. She'd simply be torn to pieces.

'So I think you'll do as I say, yes?' Camille stared at her, unblinking, and she understood that he had her trapped, like a spider under a glass. 'Where had you been by the way?'

'What?'

'Just now. Where were you?'

The anger still pulsed in her veins, but she kept her voice steady. 'The clockmaker's valet, Joseph, is trying to find a friend of his, a slave boy who worked for a furrier there. I was helping him.'

Camille let go of her wrist. 'How charming that you've made a friend here. The whore turned spy and the slave turned valet. Well, he's certainly handsome for a black.'

Her anger made her suddenly bold. 'Who is it who's taking the children? Why don't the police search for them?'

His manner changed then. He stepped away from her. 'What has this to do with what I've just instructed you to do? *Pay attention.* Concentrate on the task I've set you. It's not for you to ask questions of me.' Madeleine saw that Camille's cheeks were flushed, his eyes too bright. 'You'll report to me within a week and you'll have some useful information to tell me about Reinhart's project and about the girl. If you haven't?' He held out his hand again. Then, after a moment, he balled it into a fist.

Madeleine might have got away with it if she'd had a few more minutes to compose herself, but Véronique, returning from Lefèvre's, reached the hallway just after she did, while she was still clutching her arm. 'Madeleine? What happened? Are you injured?'

'A stupid thing. I was heating the water for tea earlier and poured some over my arm.' Even to her own ears, her voice sounded strained and false.

'Well, you must let me see. I can dress it.'

'No, no, really, it's nothing.' Madeleine felt panic spreading through her.

'Nonsense. Think of yourself as one of my first patients. Quickly, come with me.'

Madeleine followed Véronique, her heart thudding too loudly in her ears, her disordered mind trying to come up with some way out of the trap she'd constructed for herself.

Véronique led the way to her dressing room and insisted on sitting Madeleine down on her chair and lifting the sleeve to see the burn mark beneath. The girl squinted at the reddened circle of skin, then raised her eyes to Madeleine's. She knew at once, of course, that it wasn't caused by water, and Madeleine felt the heat of the burn transfer almost entirely to her face. She looked away.

The silence spread between them like a pool and Madeleine knew that if Véronique asked her anything further, she would cry or, worse, tell her the truth of who and what she was. But eventually Véronique got up, fetched some liniment and spread it gently on Madeleine's arm, then bandaged it with muslin without saying a word.

'Thank you, Miss,' Madeleine managed.

'It's all right. You'll be fine.'

Oh, but I won't, Madeleine wanted to tell her. *I won't, because my skin may heal, but my soul has been slowly burnt away by what I've been made to do, and what I'm now being made to do to you.* She stood up, smoothed her skirts. 'I've just been out looking for Victor. The little boy, remember? He's gone.'

'The boy who helped Father?'

'Yes, hasn't been seen for days. Me and Joseph, we searched everywhere we could think, but—' She shook her head.

'Madeleine, this is too much.' The blood fled from Véronique's cheeks, leaving her pale as paper.

'Yes.' And of course it was, but she hadn't expected Véronique's reaction. After all, what was he to her?

'Did you not glean any information when you went looking? Had anyone heard anything?'

'No one'd seen him. It was only the rumours about the children, wild claims of a leprous prince.'

The girl was standing up now. 'You must tell me of all the places you went, everything people said to you. You and Joseph must tell me everything you know. Then I will write to the police to insist that they find the child.'

'That's good of you, Miss. Very kind.'

Véronique was not looking at her now, but at her own white hands. 'Madeleine, it's the very least I can do. The very least thing I can do.'

15

Véronique

Be careful what you wish for. That's what Soeur Cécile had warned her once, and Véronique had thought it a stupid thing to say. Why set limits on your own dreams if they were only inside your heart? Why not aim for what you wanted rather than following the path others had set? But she thought she knew now what the woman had meant: understand how deep the waters run before you try to swim.

It had all seemed so exciting at first – to be part of the project, to be in on the secret, to have private lessons with the surgeon to the King! But then the path that had seemed so clear had warped and twisted. She'd seen, or perhaps imagined, things she hadn't wanted to see; glimpsed the rot beneath the gloss. She'd begun to doubt herself, begun to see shadows where before she'd only seen light; trap doors where there'd once been windows. Véronique moved over to her bureau and took out a sheet of paper. She hadn't dared discuss it with her father. How could she? He wasn't exactly an easy man to approach, and she had no idea what language she would use, nor how he might reply. Véronique wanted desperately to talk to Madeleine. She, surely, would know what to think and what to say. She knew, Véronique suspected, of the dark desires of men. But Madeleine, she'd begun to understand, was not her own woman either.

Véronique unscrewed her ink pot, thinking of Madeleine's dark grey eyes. There was a thread connecting them, despite the gap in age and experience and what Paris society would view as class. But she doubted that thread was strong enough to withstand whatever or whoever was pulling Madeleine away, for their grip was stronger than hers. Véronique had known that as soon as she'd seen the burn mark on Madeleine's arm, left so freshly by a man's pipe or cigar. That man, whoever he might be, was making Madeleine do something she didn't want to do and making her act against her interests. However much her maid might care for her now, she would look after her own needs first, and those of her beloved nephew. So no, she could not tell Madeleine; she couldn't tell anyone at all.

Instead, Véronique would need to conduct her own research as carefully and quietly as she could. She would begin in earnest tomorrow night. She would have to pray she was not discovered, or that if she was, it would matter little, for all her fears would turn out to be wrong. Véronique thought of Victor, the little boy with a wide smile, a silver earring in his ear. She thought of the fluttering cage of birds he'd brought her father and what they'd done to the birds inside. She thought of Clémentine and how, ultimately, she'd failed her. Her mind spanned the years to that morning when the convent bell had sounded, out of time, calling them all to the chapel. The Abbess had stood before them, her usually benign face tired and frightened, to tell of how a girl had been found that very morning, trying to climb the convent walls. 'We are all shaken,' she told them. 'But our faith in the Lord is not. In order to restore calm, to both the convent and to her soul, the girl will now do penance, as has been ordered by Pére

François. You must remember, *mes filles*, that this is for her own good. To suffer is to learn.'

The Abbess had not named the culprit but it was easy enough to work out her name, as she was the only boarder not now standing in the grey half-light of the chapel. It was the girl who had the most reason to run, but also nowhere to go.

Véronique dipped her pen in the ink pot. Should she write now? There was little she could commit to paper, but she could warn them all the same. Her pen wavered over the paper. Perhaps it was unwise to say anything without evidence to confirm her fears. She would simply be thought a foolish girl or, worse, one who'd lost her mind. Better to wait until she had something real, not something that could be dismissed as a delusion.

She caught the reflection of her own face in the window, a pale oval against the dark. She screwed the top back on her ink pot and blew the rush lamp out.

16

Madeleine

Madeleine stood in the shadows of the stairwell, the tray in her hands, her heart in her mouth, knowing Doctor Reinhart was due to leave his workroom before eleven o'clock to meet a master silversmith. It was a cheap trick, really, and reckless, but she was running out of time. Four days had passed since Camille's visit, the burn mark now reduced to a red patch of skin. Only three days remained before he returned – before he did something worse.

The clock in the hallway ticked down the seconds, the minutes stretched out like hours. She'd been spending so much time in the shadows she began to think she was more a spider than a fly, scuttling from place to place. She'd been following Véronique whenever she could, to Lefèvre's lessons at his house near Notre Dame, to the Tuileries when she crossed the bridge from the Louvre. Still, though, there were times when the girl managed to evade her – when she vanished from the house or from Lefèvre's rooms, when she returned home with her boots muddied or her skirts damp, and no ready explanation. Camille had been right: the girl was going somewhere, but Madeleine didn't yet know where. And the closer she watched her, the more she grew sure that something was tearing at Véronique with razor-sharp teeth, eating her up from inside.

Véronique had written to the police and the commissioner about Victor, as she'd said she would, but they'd done little, if they'd done anything at all. Madeleine and Joseph had been out searching again, hour after hour, but they'd found no news of the boy, no real clues, only further signs of people's mounting rage: broken timbers, messages scrawled on walls, whispers of a great rebellion.

Madeleine at last heard footsteps on the wooden floor, then the turn of the door handle. She watched in silence, barely breathing, as Reinhart emerged, locked the workroom door behind him, pocketed the key, and began to walk towards the front hall. Now! She ran softly and swiftly along the corridor, crashing into him, the tray falling, the jug of milk it contained spilling down his front.

'Oh, Monsieur, I'm so sorry!'

Reinhart was scowling, dabbing at the milk dripping from his jacket. 'Look at me!'

'Please – let me clean that for you.' She began to remove it from his shoulders.

'Very well, but quickly.'

Madeleine sped with the jacket to the kitchen, removed the key from the pocket and pressed it down into a bar of soap that she'd stolen from the last batch Edme had made. For perhaps the first time in her life, she was grateful to have grown up in the Rue Thévenot, among whores and liars and thieves. Then she cleaned the key, sponged down the jacket as best she could, and hurried back with it to the hallway.

Cursing quietly, Reinhart shrugged himself into the jacket and headed for the salon. Madeleine returned to the kitchen, wrapped up the soap and at last let out her breath. As soon as she could she'd slip out of the Louvre and head for the the crooked locksmith of the Quartier Montorgueil.

She had to wait until night-time, though her fingers itched to try the key. 'You'll need to use grease on it,' the man had told her. 'And you'll need a strong grip to turn it.' Good job, then, that she'd spent a lifetime wringing sheets, pumping water, prising lids, pulling feathers, sewing, frigging, lifting and carrying.

By ten o'clock the apartments were still, save for the whir-ring of the clocks. She crept down to the workroom, a lamp in one hand, a pot of goose fat in the other. Trembling, she removed the key from her nightdress, inserted it into the lock and turned it slowly. The damned thing jammed and stuck. Though she'd greased the lock just as the man had said, she couldn't twist it left not right, nor even pull it out. Her palms had begun to sweat and the key was now slippery as an eel in her calloused hands. Three minutes passed, four, and still the key wouldn't budge. Madeleine tried to steady her breathing, tried to remain calm, but she could feel the perspiration trick-ling along her hairline and the thudding of her heart as she imagined being found here in her nightdress, fighting with a counterfeit key. She took her hand away from the door and stood for a moment, silently praying. All right, she was hardly a good Christian girl, but He owed her this much, didn't He? She flexed her fingers, put both hands on the key and with all her might she twisted it. There was a grinding sound and then, *grâce à Dieu,* the key turned and the lock clicked open. Quietly, she opened the door.

Even in the low glow of the lamp she could tell that the room was in chaos. No wonder, given she hadn't been allowed in to clean or tidy it, no matter how much she'd asked. Piles of papers were strewn across the tables and instruments had been left out, uncovered. There was no obvious sign of any new invention, though; no mechanical monster waiting for her in the semi-dark. Quick as she could, Madeleine rifled through the cupboards, through the papers, through Doctor Reinhart's bureau. She searched for anything that might give them away. But all she could find were sketches that made little sense to her, of columns of discs, of cogs and gears and letters. A detailed drawing of a hand, with all of the tendons and bones labelled. A coil of rope, metal wire, a box full of empty glass bottles, some kind of leathery material that had been hung out on the fire to dry. Could they be keeping their invention somewhere else? Where?

And then all at once a hammering sound, as though someone was knocking on wood. Someone within the room. Her own heart pounding, Madeleine followed the sound to the cupboard where Reinhart kept his project boxes – boxes through which she'd just looked. Her hand shaking, she pulled out all of the drawers. There, at the bottom, staring at her with grotesque swivelling eyes, was a small mechanical creature, patched together from scrap pieces of metal. It was a monkey, she realised, or the prototype of one: a model on which Reinhart had been practising. The monkey held a toy wooden drum and was hitting it with a stick. Her movement had set off the mechanism, she supposed, and she willed the horrible thing to stop. It must, she guessed, be one of Reinhart's old projects, perhaps one he'd started long ago, for it had none of the intricacy or charm of his other animals that were so clearly intended for the nobility to display on a mantelpiece. This, instead, looked like a child's toy, and a fairly crude one at that. She waited until it had run itself down and shut it back away in its drawer.

★

The following afternoon, when she carried the ironed linen back to her mistress's room, Madeleine found Véronique sitting on the bed, still wearing her outdoor boots.

'You've been at your lessons, Miss Véronique? I didn't hear the carriage return.'

Véronique glanced at her. 'I decided to walk back. It's only a short way and it does me good to be out in the fresh air.'

It didn't look, however, as though it'd done her good. She looked ashen, the veins visible beneath her eyes, her mouth bruised, and she shivered.

Madeleine opened the armoire to put away the clothes. 'You're chilled, Miss. Can I get you something to warm you? Some spiced milk, perhaps?'

'No.'

'Some coffee?'

'No, thank you.' She turned her face away.

'If there's anything—'

'There's nothing.' Véronique spoke abruptly and Madeleine felt a jolt of rejection, then the cold spread of unease. She'd wondered if Véronique might've been going to a lover – one of those satin-clad boys – but she didn't look like a girl come rose-cheeked from a rendez-vous. She looked, in fact, like poor Suzette after a visit from Camille. Silently, she unlaced Véronique's boots and took them away, leaving her sitting on the bed, staring at the wall. When Madeleine crept up the stairs a few minutes later, she saw that Véronique was still sitting in the same place, with her knees pulled up to her chin and her arms wrapped around her. Madeleine stood there for a moment, watching her, wanting to go to her, to hug her. But she found that she could not.

She didn't see Véronique again until the evening when she prepared for bed, and even then she was quiet and withdrawn, her eyes very large in her bone-pale face, her fingers worrying her mouth. Madeleine buttoned her nightdress for her in silence, the tiny pearl buttons iridescent in the candlelight. 'Will you not tell me what bothers you?' she asked at last. 'Whatever it is, I'll keep it to myself.' Lies of course, yet more lies, but she wanted it to be true.

Véronique watched her for a time in the looking glass, and Madeleine thought she was about to say something, but she merely shook her head. 'You needn't worry about me, Madeleine. I've spent all afternoon helping my father and Lefèvre and too many thoughts are whirling about in my mind.'

Madeleine began plaiting the girl's hair, as she'd become fond of doing, and for a time they were silent. Could it be, Madeleine wondered, that the King himself had been pawing at her mistress? 'If someone is making you do something you don't want to do—' She stopped. What advice could she possibly give when her own life had been largely spent being forced to do things she hated?

'I am doing what I have to do, Madeleine. Just as we all are. Getting on with the job in hand.'

'What is it, Véronique, that you're making with your father? Can you not tell me anything at all?'

A mild shake of the head again. 'I can't. I swore I wouldn't.' Véronique toyed with a spun glass bottle on the dressing table. 'You wouldn't like it in any event, Madeleine. You didn't even like that doll I made.'

'It was only the eyes,' Madeleine said quickly, 'that unnerved me a little. The clever way they moved. It made me feel like the doll was alive, somehow.' And it evoked memories she'd believed long dead, but she couldn't tell the girl of that.

Véronique met her gaze in the glass again. 'Yes, it's about the line between life and not life, and perhaps you were right to be bothered.' She looked away. 'Now I must go to bed.'

'Is there anything I can get for you? Anything I can do?'

Véronique shook her head.

Without thinking, Madeleine squeezed her shoulders as she would once have done Suzette's. 'Very well, Miss. But I'm just next door, and I'm here if there's anything you need. You know that, don't you?'

Véronique turned to her and Madeleine could see the tears shimmering in her amber-green eyes. She gave a brief nod. 'Thank you, Madeleine. But honestly, there's nothing you can do.'

Slowly, Madeleine descended the stairs to the kitchens. She had a strong feeling that she shouldn't leave Véronique, but the girl had told her to go. She thought of how Suzette had been at the end – sparrow shoulders and a swollen belly. *Don't leave me, Madou. I'm scared on my own.* As if she'd have ever left her, but small comfort she had been. Suzette stayed in Madeleine's mind as she scrubbed the pots in the scullery, scouring them with sand as she remembered Maman's refusal to go for a proper physician, her insistence it would all be fine. So immersed was Madeleine in her work and her thoughts that when the small boy appeared before her she nearly shrieked. It was the ashen-faced child who'd first brought the note from Camille. In the gloom of the kitchen he could almost have been a ghost.

'How d'you get in?' she muttered, putting down the pot she'd been scrubbing and drying her swollen hands.

The boy didn't answer, but held out a letter, then waited as Madeleine fished around in her petticoat pocket and brought out a single dull coin. 'He said you was to burn it, then to come at once.' He took the coin and ran, his footsteps barely audible as he moved into the hallway and towards the outer door.

Madeleine cracked the wax on the letter and read the short note inside: *'You are to come to the Châtelet immediately. Say you are here to see Inspecteur Viconte. Ensure no one knows where you are.'* She abandoned the washing, threw the note into the grate and hurried into the hall to collect her cloak, her heart kicking against her ribs. The Châtelet. It was where the police headquarters were located, where the very worst criminals were kept. What did they want with her there? What had she done wrong?

She left as quietly as she could, her shadow flitting behind her as she descended to the front hall. Outside the sky was swirled ink sliced through by a shard of moon. Looking up at their apartments from the Louvre steps, she thought for a split second that she saw a face at the window, but it was, she decided, simply the play of the streetlamps upon the glass.

On the Quai de la Mégisserie carriages made their way to the playhouses. Workmen, carpenters and masons walked back home from the inns, still caked in plaster, soot or paint. A cold wind blew across the water, skewing the candles in their lanterns or extinguishing them entirely, so that some patches of Paris were plunged into blackness, others illumined only by a wavering glow. More than once Madeleine grew convinced she was being followed, but when she turned, there was no one there – only her own reflection sliding past her in darkened windows, only the echo of her footsteps on the ground.

The area around the Châtelet was a labyrinth of ever narrower and dirtier alleys where butchers' shops clustered and curdled blood pooled in the darkness of corners and doorways. Madeleine paid three sous to a link boy, who hurried before her, through the stench of the tallow-houses, the flame of his torch bleeding into the night. Within ten minutes they'd reached the archway of the Châtelet compound and, as Madeleine walked beneath it, the blackened facade of the Châtelet grew before her: a medieval fortress, a castle from a nightmare, flanked by towers like protruding teeth. She was approached by a man in dark clothing who gave off an odour of gin and damp wool. When she uttered the name written on the piece of paper, he led her through the iron gates and along a long grey wall, the windows merely slits. She imagined prisoners chained to the other side of that wall in tiny cells or torture chambers, their screams absorbed by the grey brick, their faces forgotten outside. She followed the man up an evil-smelling staircase to an office where clerics were writing by candlelight. From there she was taken up to another floor and told to knock three times upon the oaken door.

By this stage, Madeleine's breathing was ragged, her face was sticky with sweat. She took a moment to calm herself, listening to the murmur of voices within. Among the men's voices was one higher pitched: a woman's. She waited a moment longer before knocking gently on the door.

Within a heartbeat the door swung back. A valet in dark grey stood before her, assessed her, then moved to the side to allow her to enter. Madeleine saw Camille leaning back on an armchair, Berryer seated behind a large desk and a woman, her back towards her, wearing a dress of emerald green. Even before the woman turned, Madeleine knew who she was, yet she almost gasped to see her face. She was thinner than when Madeleine had seen her in the shop and her face had been painted thickly with white, the red spots on her cheeks making her look like some exquisite doll.

'Come forward,' Berryer said.

Madeleine walked closer to them and curtsied. She kept her head down, but watched Pompadour through her lashes. What the devil was she doing here in the squalor and stink of the Châtelet?

Berryer moved some papers to the side of his desk. 'Miss

Chastel, we wish to know what you've discovered. Tell us what the clockmaker is making.'

Madeleine felt her blood chill. Hadn't Camille said she was to have a week? 'I still don't know, Monseigneur. I managed to get into his workroom, but it wasn't there. He must be keeping his invention somewhere else.'

'You must have some idea what it is.'

She thought of the wax doll, the sketches, the tray of eyes. She thought of the vial of red liquid. 'I couldn't say.' Because she might very well be wrong.

Pompadour beckoned her to come further forward. Even standing a foot away from the woman, Madeleine could smell her scent – powerful, exotic, expensive.

'You're an intelligent girl. You can read. You must have some idea of what they're about, of what they've been asked to do.'

Madeleine hesitated and Pompadour curled her perfect lips into a smile. 'Louis must seem to you to be a very powerful man, but he's also vulnerable. He has morbid fascinations, fascinations which do him no good. He was brought up with death, and it continues to occupy him.'

She guessed that Pompadour was talking of his parents and his siblings dying when he was tiny, about how he himself nearly died of smallpox not long after. She'd been told about it when she was young and had felt sorry for him – this man who'd been made King at the age of only five. But then, aged twelve, her own childhood had been taken from her. She didn't feel sorry for him now.

'I didn't hear what the King asked Doctor Reinhart to do. I haven't seen any instructions either. But ... Miss Véronique told me that I wouldn't like it.'

'She's helping him, then.'

'Yes.'

'How is she helping him?'

'I don't know.'

'Then what, exactly, have you been *doing*?' Camille asked.

Pompadour put up her hand to silence him. 'Madeleine – that's your name, isn't it? I am very good at telling when people are lying to me, and when they are holding something back. What do you think they are making?'

Madeleine swallowed. 'I swear I don't know for sure, Madame. But I fear that it's something ... perverse perhaps. Against nature.'

'You think they are making a human.'

Madeleine pictured very clearly the material drying on the table, the sketch of a human hand. 'Perhaps.'

'Well, you must do your best to find out. You must establish what they intend to do. I am rather worried about it all, you see.'

'Yes, Madame,' she said. 'I am doing my best.'

Pompadour looked closely at Madeleine's face as though it were a book to be read, then turned back to Berryer. 'The clockmaker has sent no correspondence that would help us?'

Berryer shook his head. 'Marquise, I would tell you immediately if there were anything of interest to you.' His tone with Pompadour was entirely different to that he'd used with Madeleine: gentle, almost simpering. Madeleine guessed this was the way that all men spoke to her, save, perhaps for the King.

Camille was staring at Madeleine. 'What have you learnt about the girl?'

Berryer shook his head at him.

'Nothing, really.' Or at least nothing that she would tell them.

'Then things are not looking very good for you, Madeleine.'

She felt her heart pulse faster, felt herself grow sick. She stared at Camille's grey-tinged skin, the pale eyebrows, the thin neck. How she hated him. How she hated what he'd made her do, what he'd made her sister do before her. 'Why are you asking me—'

'It's not for you to question your duties,' Berryer said sharply. 'You are simply to carry them out. You must keep the household under constant scrutiny. You must continue to report back regularly on anything of interest that you find. You know, I presume, what happens to *mouches* who refuse to do their duties.'

She licked her lips. Oh yes, she knew all right. They were squashed, quickly or slowly, depending on the inclination of the person doing the crushing.

'Please, Monseigneur: my money that I was promised. I still haven't been paid a sou.'

Berryer raised his eyebrows as if this request astonished him. 'You dare to ask for money when you haven't yet completed the task my officer set out for you? When you haven't yet determined what Reinhart is constructing?'

'My nephew, Émile, I need to return to him, to help him. And Camille told me—'

'You will be paid when you have done what you've been asked to do. Now you may go.' Berryer pointed at the door. 'But remember, that wherever you are, we are there too.'

Madeleine gave a brief, mechanical curtsey, though she wanted to spit in his white painted face, though she wanted to be shot of the lot of them.

As she left the room, she sneaked one last look at Pompadour. She was leaning forward, towards Berryer, and his hand was almost touching hers. 'It's not Louis' fault,' she was saying softly. 'He thinks he's just playing. This is what comes of not giving children any toys.'

★

It was gone ten o'clock by the time Madeleine reached the Louvre, frozen through and trembling so violently she could scarce turn the key in the door. Her mind raced with how she might tear herself from the police's grip, how she might get what she was owed. She needed to get Émile out of the brothel and she couldn't stand to be trapped any longer, to be made to do things she hated, to betray the only girl she'd ever really cared for other than Suzette. She was so focused on her own predicament, so chilled by the night air, that she failed to see her until she almost walked into her, standing motionless in the hallway.

'Edme! What a turn you gave me.'

'Where is she?'

'Who?'

'Demoiselle Véronique. Where is she? Where have you been?'

Madeleine had prepared an answer to the last question. 'I've been to my family. I received word my mother had taken a turn for the worse.'

'Véronique wasn't with you?'

'No, of course not. Why would she be?'

'She's gone. Gone from her room and no one knows where.'

Madeleine's heart seemed to dip. 'She was here but two hours ago. I helped her dress for bed.'

'Well, maybe you did, but there's no sign of her now and her cloak and hat are missing. Her father noticed the light in her room had been snuffed out at not a quarter past nine and got no reply when he knocked on the door. The counterpane was drawn up and no candle had been lit.'

'Where are the others?'

'Doctor Reinhart has gone out to look for her, Joseph too, and I've been left here in case she returns, and every time I hear a sound I think it's her, but no.'

209

Madeleine took off her hat. 'Véronique has gone out on her own before.'

'Oh, I know that, girl, but she's never been out in the evening before, has she? She shouldn't be out in this city, not at this time. Not with them taking the children.'

'Tradesmen's children, Edme. Twelve and thirteen-year-olds. They wouldn't take a young lady.' Madeleine's voice faltered as she heard the ridiculousness of her own words. She had no real idea what was going on, or who might be at risk. In any event, in her plain cloak, Véronique might well have looked like the daughter of a chandler or baker. She could, in truth, have been anyone.

'She looks younger than her seventeen years, Madeleine. You know that. And in the darkness ... *Cher Dieu*, why on earth did she go out at night?'

They went together into the salon and sat for a few minutes without speaking, listening to the sounds of the city outside. Then came a bang as the apartment door swung open and hit the wall. Doctor Reinhart stood in the hallway, the rain glittering on his coat. 'You. Where were you? What have you done with her?'

'Nothing, Doctor Reinhart.' Madeleine's heartbeat was like a roar in her ears. 'I was at my mother's. I thought Demoiselle Véronique was here.'

He walked closer to her, looming over her, his face half in shadow, the cold of the night rising from his coat. 'She must have told you she meant to go out.'

'No! I helped her into her nightgown.'

'Did she seem upset? Distraught?'

Madeleine hesitated. 'I was only with her for a few minutes. She was preparing for bed, not to leave the house. She didn't seem upset, exactly. Quiet, though. Worried.' As she had been for days. Had he not noticed that himself?

Doctor Reinhart's eyes were still on her, very large and very bright. For a moment she thought that he knew it all: every lie, every trick, every betrayal. 'If I find that you've lied to me about this, or that you've hidden something from me, it will go very badly for you.'

Madeleine retreated backwards, towards the wall, shivering. 'I don't know where she is. I promise you that.'

Edme put her hand on her master's arm. 'Now, Doctor Reinhart, don't be scaring the girl. There'll be an explanation for this and we'll find it. Let me get you something to drink. A hot posset, or a little plum brandy. Something to calm your nerves.'

He pulled away from her. 'There's nothing wrong with my nerves. My senses do not need to be dulled. On the contrary, they need to be sharper than ever. I need to work out where she is.'

With that, he marched down the corridor, unlocked his workroom door and then they heard the closing of the door, the scrape of the key and then a thud, as if perhaps he had thrown something, or fallen to the ground.

Not knowing what else to do with herself, Madeleine went up to Véronique's dressing room, to look for some clue as to where she might have gone. She lit the candelabra and as the flames reared up she saw for an instant Véronique's face, staring at her from the mirror. Madeleine blinked and the vision was gone – a horrible trick of the mind. She must try to calm herself. In the candles' silvery glow, she collected the brushes and combs, the spilt pins and ribbons; she found a stray button, a jewelled pin, a ball of paper on which she had dabbed vermilion from her lips. Had that been from this evening? Had she put on paint before she left?

She opened the dresser drawers, releasing the scent of lavender and dried rose petals. She patted down her small clothes and slips, the silk placed in paper. She took out the stockings and gloves, looking for some new item or gift from a lover that might give her away. There was nothing, however; nothing she hadn't seen before, hadn't washed, dried and folded with her own roughened hands. Véronique kept her few jewels in a locked box on the windowsill. When Madeleine picked it up, she could feel the movement of the jewellery within.

Madeleine opened Véronique's armoire and checked her dresses. The rose silk taffeta was there, and the green silk, the white cotton, the pale silver, but the mauve one was gone. She must have dressed herself again after Madeleine had prepared her for bed. Why? She felt suddenly cold. Had she done this before and Madeleine not known, going through the pretence of readying herself for bed, when in fact she was intending to go out, like one of the twelve dancing princesses?

Madeleine went to Véronique's bed, lifted the counterpane, and ran her hand beneath the mattress. Nothing. Finally, she looked beneath the bed and her breath caught as she saw a tiny face peering at her from the shadows, but it was only a doll. She brought it into the glow of the candlelight, remembering again with a twist of sickness the old man of her youth and his porcelain doll. This was no shiny new gift, mind, but the strange toy that had once sat on Véronique's mantelpiece, the paint worn from its cheeks, its clothes ragged with age.

'*You wouldn't like it, Madeleine,*' Véronique had said about the project. '*You didn't even like that doll I made. I'm not sure that I much like it myself.*'

17

Madeleine

Madeleine slept poorly and woke early, the pale light oozing beneath the curtain, but when she went downstairs she realised Reinhart had already left.

'He didn't go to bed,' Joseph told her when he emerged not long after, his skin dull with fatigue. 'I could hear him pacing in his room at gone two o'clock. He said he would go again to the police this morning.'

Edme was already at work stirring coffee in a pan. 'Why did she go out? That's what I want to know. Where could she have been going?'

Where indeed. 'We should go to the hospitals,' Madeleine said. 'In case she's been injured in some way. Maybe start with the Hôtel-Dieu and La Pitié. We should try the prisons too.'

Edme began pouring the coffee into cups. 'What on earth would she be doing in a prison? Why would they have arrested an innocent girl like her?'

Madeleine raised her head to look at Edme. What an outlook on life to have: to assume that you could only be locked up, only be harmed by the state, if you'd committed some kind of crime. 'I'll go to the Bastille, then the For-l'Évêque. Joseph, you try the Hôtel-Dieu. We will meet back here at midday.'

Paris had never smelt so rank, even in the height of summer. There was something putrid in the air, intensifying as Madeleine walked further along the Rue Saint Antoine, as though the pestilential miasmas of all the overfilled graveyards and squalid basements were seeping out to sicken the city. And then she caught sight of it. Madeleine had often seen the grey hulk of the Bastille, but never had cause or urge to approach it, for it was a place despised across Paris, a place where men were buried in stone. As she neared the drawbridge, the walls and turrets seemed to rear in the sky, obscuring the watery sun. In the gloomy courtyard, servants and carters were carrying supplies across the straw-strewn cobbles, and soldiers shouted at a line of people: mainly women in dirty clothing, some with tiny children strapped to them, some carrying bundles of food or clothing, all trying to gain access to the prison.

Madeleine joined the line and waited, watching the sparrows squabbling over dropped crumbs of bread. A terrible thought was snaking its way through her mind: could Véronique have followed her to the Châtelet last night? Had she worked out she was a spy? The fear was like a hook in her flesh.

'I tell you again, Monsieur,' she heard one of the soldiers bark at a man in a stained leather apron, 'your son is not here. Just as he was not here yesterday, or the day before.'

Madeleine straightened, her eyes sharpened. The man in the apron had a wild look about him, his face haggard, his clothes filthy. 'But this is what some people are saying now: that a boy was brought here and kept for a week. My neighbour—'

'Your neighbour doesn't know what he's talking about. There are no children in here.'

Madeleine, ignoring the mutterings of the women before her, stepped forward to speak to the soldier. 'Has a young lady been brought here, though? In the past day? Véronique Reinhart. She's my mistress.'

The soldier cast her a brief look. 'Unless the police have told you to come here, you're wasting everyone's time.'

'But—'

'No young madam would have been brought here,' the second soldier told her. 'What d'you think this is? A bloody dancing hall?'

'She wouldn't have looked like an aristo,' Madeleine persisted. 'She would've been wearing a plain grey cape like any young grisette.'

The soldier waved her away and moved on to speak with another woman who was begging to see her husband. Madeleine gave up on the soldiers and hurried instead after the man in the leather apron, who was now approaching the outer gate. 'Monsieur, wait!'

Close up she could see that his skin was patched red in places, as though he'd been at it with a nutmeg grater, though more likely he'd been at the bottle. 'I heard you speaking with that man. You're looking for your son, are you?'

'You know something?'

'I know nothing, I'm afraid. My mistress has disappeared and I'm trying to find out where she might be. I heard you say you'd been told a boy was kept at the Bastille for a week.'

'That's what my neighbour told me, who was told by a friend of a friend.'

Madeleine nodded slowly. Children were sometimes arrested for vagrancy or theft, so this was no great surprise, but the police would've had to have gone to a commissioner. The families would've been told at once. 'When did your son go missing?'

'Philippe is his name. We lost him at the end of January.'

January. Over two months ago. No wonder the man looked

wild. No wonder he'd taken to drink. 'Where did he go missing from, your Philippe?'

'Well, the last time he was seen it was where he worked, on the Place Dauphine.'

A blade of fear cut through her. She remembered her very first conversation with Edme, the day she'd come to the clockmaker's house. 'Your son was the baker's boy,' she said quietly.

'You heard about him, then.'

She'd heard but she hadn't listened. 'And this other boy you mentioned, who was supposed to have been kept here for a week, where did he disappear from, Monsieur?'

'I don't know. I know very little about it so far. Now you'll excuse me, Demoiselle, but I must search the other prisons. I'm not accepting that there's no hope.'

*

The day passed as a series of broken fragments, Joseph and Madeleine both returning to the Louvre without news of Véronique, both setting out again to search further. Lefèvre arrived at the apartments late in the afternoon, while Doctor Reinhart was still out. He seemed shaken, devoid of his usual charm, and when Madeleine offered him some coffee and marzipan tart, he said only: 'How long has she been gone?'

'Since last night, Monsieur. About nine o'clock. Maybe earlier.'

Lefèvre sat heavily on the velvet settee, his face sagging. 'Why did Max not get word to me at once? He knows what influence I have.' He shook his head. 'Well, I'll do what I can to ensure that the police search properly for her, that Louis directs them to do so. I will go to Versailles at once.' He looked at Madeleine for a long moment. 'Would you say that your mistress was content, my girl?'

216

The question caught her off guard. 'I don't quite take your meaning, Monsieur.'

'Certainly you do. You were a companion to her as much as a maid, I saw that. You must have had some notion as to her state of mind. Did she say anything to you to suggest something was worrying her?'

Madeleine looked at the floor. After a moment she said, 'Véronique did seem a little ... preoccupied these past few days. Like there was something eating away at her.'

'Have you any idea what?'

Had she? She'd wondered then if it was to do with the invention. Then she'd wondered if it was a man. The King. She wondered now if it was because Véronique had suspected her, or found her out. But of course she couldn't say any of that. 'No, Monsieur, I'm afraid I don't. She kept quite dark about it.'

'Indeed?' He seemed thoughtful. 'Now that I think about it, I agree she has seemed a little distant these past two or three days, her mind not entirely on her lessons. I should have said something.' He looked at her again. 'Can you think of anywhere she might have gone? Did she have some friend she might have decided to visit? Someone from the convent, perhaps?'

'No, Monsieur.'

'No one else to whom she might have told her troubles?'

Madeleine shook her head. 'So far as I know, she has no friends in Paris.' *Except for me, and what kind of friend have I been?*

'I see. Poor child.' He stood up. 'No matter. We will find young Véronique, wherever she has gone, and we will make things right again, won't we?'

Madeleine nodded as was required of her, but she saw through his bluster to the anxiety beneath, for he pulled absently at his lip.

'Has she had any callers of late? Any gifts, that sort of thing?'

Madeleine felt a twist in her stomach. 'Not that I know of, Monsieur, but then she might not have told me.'

'You have … looked through her room? Not found any item, any papers that might give a clue as to her whereabouts?'

Madeleine watched him. She did not want him to know her for a spy, but nor did she wish to seem a fool. 'I've looked thoroughly, Monsieur, and there's nothing. Had there been, I would have told.'

Lefèvre nodded and touched her arm lightly, at last attempting a smile. 'Yes, yes. You're a good girl. Leave it to me now. I'll do everything I can.'

Late that evening, as rain battered against the windows, a man came, smartly dressed in black. He insisted on speaking to Doctor Reinhart at once, alone. When Reinhart reappeared in the hallway he was wearing his coat and his eyes were as dull as pebbles. 'There is a girl's body in the *Basse-Geôle* mortuary,' he told Joseph and Madeleine, 'a girl with fair hair. I must go to there at once. I must find out if it is her.'

Madeleine felt darkness at the edges of her consciousness, felt the fear that had been holding her upright threaten to give way to something else.

'I will come with you,' Joseph attempted, but Reinhart shook his head.

'No, Joseph, it isn't something I want you to see. It is not something I wish to see myself.'

But it might not be her, Madeleine wanted to shout. It might just be some other girl! For weren't there always bodies being dragged out of the Seine, or found frozen or starved or beaten in doorways?

Joseph passed Reinhart his hat and cane and then opened the door for his master. Reinhart nodded, his face a blank, and then walked out into the approaching night.

Reinhart was gone for what seemed like an eternity, but was, according to the clocks, little more than an hour. They were all assembled in the hallway when he came in, the rain dripping from his hat. He kept his head down at first so that Madeleine couldn't see his expression, but from the way he stood, back bent, hand clutched around his cane, she knew what he'd seen in the *Basse-Geôle*. She knew all hope was gone.

Joseph stepped forward and removed his master's wet coat, then his hat. For a moment, Doctor Reinhart removed his glasses and she saw that the skin behind them was thin and pale like the wing of a moth, his eyes a watery grey.

'It was her,' he said at last. 'She is dead.'

None of them moved. The only sound was the ticking of the clocks, distant shouting from the street outside. Edme covered her face with her apron and began, very quietly, to cry.

'How?' Madeleine asked at last.

He wiped his brow. 'According to the stretcher-bearers who brought her to the mortuary, a carriage ran into her, yesterday evening, on the Place Baudoyer.'

'Whose carriage?'

'It isn't known. A man in a single-horse chaise. The driver didn't stop, which of course is nothing new.'

'But surely someone must have seen?' Madeleine knew she was asking too many questions for a servant, that she should leave the man alone, but she had to know, she had to know everything, had to know if it was her fault.

'If they did, they have not come forward. I will ask the police to find witnesses, but what I don't understand is why she was out there; why—' His body seemed to tilt and Joseph put his hand on his master's shoulder to steady him.

'Enough questions now, Madeleine.' To Edme he said, 'Some brandy.'

Edme collected herself, wiping her face on her apron. 'Yes, of course. Master Reinhart, you get yourself upstairs and dry. I'll bring some brandy and biscuits.'

'The clocks,' he said, seeming not to hear her. 'We need to stop the clocks.'

'Which clocks, sir?'

'All of the clocks,' Reinhart said, too loudly. 'We must stop all of the clocks. She is dead!'

He and Joseph divided the task between them, Reinhart taking the hallway, the salon and his workroom, Joseph the parlour and the upstairs rooms. Gold clocks, silver clocks, watches and fobs, all cut off mid-tick. Madeleine, meanwhile, was tasked with closing all the shutters so that the house was swallowed in gloom. At first, the repetitious nature of the task, and her habitual tendency to push her feelings down, kept her going, but by the time she reached the salon it was though a wave was crushing her beneath the sea, as though her chest would burst. That beautiful girl full of life and determination – that girl who'd treated her as a real person despite her scarring, her lowness, her lack of learning – was gone. And somehow the deadness that Madeleine had carried in her since Suzette died, that coating of numbness that had protected her for months, was torn up, exposing the nerves beneath, sending searing pain through her, such pain that she thought would break her. She sank to the ground on the salon floor, remembering how Véronique had lain here stretched out, her feet bare, asking Madeleine about their destinies.

After a time, an arm went round her and pulled her close. Joseph was kneeling beside Madeleine, holding her. He didn't say anything, because what was there now to say?

When all the clocks and watches were stopped, and when Madeleine had dried her swollen face, they collected in the parlour where Doctor Reinhart sat, upright and still, his eyes vacant.

'It is not natural,' he said to them. 'It is not the natural order of things. The natural order of things is broken.'

'Master, it is late. You should retire to bed now,' Joseph said.

'Yes. Very well.'

As she watched Reinhart walk slowly from the room, Madeleine noticed that his hems had come down, that his stockings were splattered in mud. For a moment she and Edme stayed in the room, unspeaking. The place was peculiar in the silence. It was as if the heartbeats of the house had stopped; as though the building itself were dead.

★

Word of the tragedy spread quickly through the veins of Paris, as macabre or salacious news always did. White flowers and black silk arrived, unwanted visitors, ostentatious gifts and letters of sympathy from clockmakers, courtiers and flatterers. Officers of the Watch came too, dark-suited and solemn, but with no news of any witnesses. In the streets, rumours rustled of the missing children, and anger fermented further. People gathered on corners, in wineshops and halls to talk, to vent, to plan.

Madeleine kept herself busy, assisting Edme with ordering the black wool and silk in which Doctor Reinhart would now dress, and the mourning bands that they must now all wear. During the day, she went about her tasks numbly, mechanically, but at night, grief stole up upon her like a phantom. Madeleine had always prided herself on her resilience. She'd always managed to push things down or away. They didn't always stay down, of course – now and again she'd be gripped by a sudden stifling feeling, a tightness in her chest, which could be brought on by any number of things: a smell, a touch, a porcelain doll – but mostly they were out of her mind, or at least pressed into its darkest corners.

But with Véronique's death, something strange happened. At night, as if from the bottom of a well, memories clawed their way to the surface. She pictured again and again not just Véronique's death, but Suzette's. Remembered her screams, her pleading, then the slow leaching of life as the baby refused to come out of her too-young body. The midwife's uselessness, her own helplessness in that hot and airless room. When she tried to remember Véronique's face she often saw her little sister's instead, pallid and swollen, clammy with sweat, tendrils of hair plastered to her forehead. 'Don't leave me, Madou,' she'd said. And Madeleine hadn't. She stayed with her for two whole days, two days that felt like a lifetime, two days that had not only killed her sister but also drained something from herself. So that when finally they accepted there was no heartbeat in either mother or child, Madeleine's own heart seemed to have dulled. And that had been all right, because the important thing was to fulfil her promise to her sister: she would get Émile away from there. She'd give him the childhood they'd been made to forfeit. That was why she must now pick herself up, dust herself down, and rebuild the wall she'd built around her.

On the day before the funeral, their royal visitor returned. He came alone and wore a dark brown cloak, as though he were any ordinary cove. Camille had told her about that a long time since: how Louis liked to go out dressed as a bourgeois and rub shoulders with the man in the street, or, more likely, the woman. As Madeleine took the coat from his shoulders to reveal the mulberry velvet beneath, it occurred to her that he was playing a part just as much as she was – had done his whole life, most likely. What role had he played for Véronique? Was it him who'd made her unhappy? She served him his coffee, as she'd done before, and set down a plate of sugared figs, then waited outside the parlour door to listen.

'I come to offer to my condolences, Doctor Reinhart. Such a loss. Such a waste.'

'Thank you, Majesté. With all the arrangements, I'm afraid your project …'

'You need not concern yourself about that. It is important, of course, that we achieve our aims, but you must have your time to grieve.'

'Yes.'

'It was a carriage accident?'

'That is what I was told, yes.'

'You saw her?'

'I did.'

'What did she … how did she look?'

'Look?'

'Yes. In death. Did she seem peaceful? Or otherwise?'

For a while there was silence. Madeleine imagined Doctor Reinhart floundering for words. What kind of a question was this?

'Peaceful, no. Her face … Her body, twisted …'

Madeleine imagined, without wanting to, what Véronique would have looked like after being trampled beneath a horse. She closed her eyes against it.

'But she was still beautiful,' Louis insisted.

'Well, you could tell what she had been, but she was imperfect. Damaged.'

Another silence.

'I was five when my great-grandfather died,' Louis said. 'They took me into his chambers so that he could speak to me, his heir, before the end. He told me, I remember, of my

223

obligations. My duties to the country, to God. And I have fulfilled them, of course,' he added in a rush, then paused. 'The smell, of course, was very bad, what with the gangrene being so advanced, but there was still a great dignity to him even at the point of death. Still beauty. That is how I choose to remember him.'

'Yes, I'm sure.'

'Did you know that after I die, my heart will be preserved, just as his was?'

A pause, Reinhart no doubt wondering what on earth to say. 'I confess I had not thought of it.'

There came a rustle of fabric as the King stood up. 'I would like to do something to preserve Véronique, to mark her memory, her life. I was thinking ... a fountain in the Place Dauphine. For her vivacity. Her sparkle.'

'That is very good of you, Majesté.'

'I normally leave the plans for such things to the Marquise de Pompadour, but I think in this case that would be impolitic. I will speak to an architect – have a design drawn up.'

'Yes, thank you.'

Footsteps moved towards the door at which Madeleine stood. She moved back.

'I will send word,' Doctor Reinhart said, 'when I have managed what we discussed. I hope that I can do it.'

'I do not doubt that you can, you of all men. But for now you must grieve. There will be time for our project. It is perhaps even more important now than ever.'

'Yes. Indeed, it occurs to me, sire, that it should be dedicated to Véronique, her having been a vital part of it.'

For a moment, there was silence. Then the King said, 'Yes, I approve of this idea, Reinhart. Far better than a simple fountain. Perhaps we could name it after her? And now I will leave you in peace.'

A woman had been brought in to prepare the body – a woman so bent and drained by age that she was surely not far off being a corpse herself. Madeleine had given her Véronique's white dress and her best white gloves, some silk stockings and her embroidered shoes. 'Thank you, *chérie*. We will make her look fine. You'll barely know she is dead.'

Perhaps she shouldn't have looked, but Madeleine had to see her for herself. That night, once Reinhart had finally gone to bed, she entered the salon where the body was laid out. Only a rush-lamp was lit, casting strange shadows about the walls, and incense was burning in the wall cavities to obscure the smell of decay. She took the seat by the coffin, which had been left open, white flowers arranged around the body, their smell sickly sweet. The woman had certainly spent her time on Véronique. The girl's face was now so thickly painted you could barely see the skin beneath, could see no sign of the damage that Reinhart had spoken of, only a gloss of white, with the lips painted a deep pink, the cheeks given a flush of rose. She'd been made to look like a court lady, though she'd never be presented now. So little time she'd had; so little of the life she so much wanted.

Madeleine felt a pain in her throat and looked away from the face, down to the gloved hands that were folded over her chest, over a gilded book. It was, she saw, the book of automata Véronique had treasured – the one she'd kept by her bed.

Her eyes burning, Madeleine silently said farewell. *You deserved a deal better than this life gave you. You deserved a deal better from me.*

18

Jeanne

'Is he unhappy in there, Maman?' asked Alexandrine.

Suppressing a cough, Jeanne peered through the bars at the mountain lion pacing up and down in his enclosure. He had only been here a month and already his golden coat had lost its lustre. 'I imagine he'll get used to it, *mon tresor*.'

'I don't think I'll ever get used to the convent.'

'Of course you will. They'll do everything they can to make you comfortable. It's simply a matter of time.' And of adapting yourself, steeling yourself. A pity Alexandrine had had to learn it early, but there it was. Louis had been intractable on the issue and now was not the time to argue with him. She needed to pick her moment.

'I don't want to go back there, Maman. It's cold all the time. And I don't like the food – it's *dégueulasse*.'

'Well, let me talk to them. I can have some sweet things brought to you. Some sugar almonds, some jellies, some fruit from the gardens here. Should you like that?'

A tear was sliding down her daughter's cheek, which Jeanne wiped away with her fingers. Only occasionally did she question if she had made the right choices, and it seemed one of those moments was now.

'Come. Let's go and find the zebras, shall we?' She took Alexandrine's little hand and led her down one of the paths leading towards the central pavilion. She regretted bringing her daughter here. The menagerie had been neglected over the winter and the place had a forlorn look to it, ostriches scratching at the bare ground, the wolves half starved, exotic birds sitting miserable and silent, their magnificent wings clipped. It had, people said, been a glorious place at the time of Louis XIV, attracting visitors from throughout the land. Her Louis, though, had little interest in the animals, save for when they were dead, only keeping the place running because he couldn't possibly let lapse something his legendary great-grandfather had initiated. The same went for the rigid and ridiculous regimen that the Sun King had created; her Louis could never abandon it, even though it made his life a misery. Staring at the caged birds, Jeanne thought the whole enterprise as cruel and pointless as the rules of court. She would encourage him to close it down.

As they reached the pavilion she saw that two people were approaching from another path. Her breath caught in her throat as she realised it was Richelieu, his movements as smooth as those of the wolves, a painted duchess on his arm.

'Marquise.' He nodded at her, though etiquette dictated he should bow. 'Brought your daughter to meet the beasts, have you?'

The woman next to him gave a smile: pretty, practised, false. 'Your daughter? How delightful. How very like you she is.'

Jeanne pulled Alexandrine closer to her, not wanting them even to look upon her. The woman's chest was flushed, Pompadour noticed, her cheeks too, beneath the paste. Clearly they'd been rutting like beasts themselves somewhere in the palace gardens.

'Indeed,' Richelieu agreed, 'and I must say you yourself look younger every day, Marquise. However do you do it?'

It was him. She was sure of it, then. He who'd left the *poissonades*. He who poured poison in Louis' ear while pretending to be syllabub sweet.

'You are too good, Monsignor, truly. But I regret we must leave you – we are late for an appointment elsewhere.'

'What a pity,' Richelieu said. 'We have so much to discuss. Terrible news about the clockmaker, wasn't it?'

She froze momentarily.

'Oh, did you not hear? His daughter died. I only know because Louis has asked Henri to design a fountain in her honour. He didn't mention it to you?' He waited a beat before he said, 'I wonder why not. But perhaps you haven't seen him much of late.'

'Well, you know how it is, Monsieur. He may be the King, but he is not the master of his own time. Now you must forgive us, but we are due elsewhere.'

Jeanne retraced her steps more quickly along the path, knowing that their eyes followed her. She gripped her daughter's hand even tighter. In the distance a peacock screamed.

*

When Berryer came to her, Jeanne was on the roof of Versailles, feeding the King's hens. She always found them oddly calming. Reassuringly predictable in a place that was anything but. In their cages, the pigeons cooed.

He swept into his usual bow. 'Marquise. You called for me.'

'Richelieu told me the clockmaker's daughter was dead. Tell me what you know; tell me what is being said.'

'Yes, a most regrettable affair. Yet another carriage accident. You may be aware of how common they are.' Berryer shrugged his shoulders. 'This is the problem with overpopulation, you see. Too many people, too many carriages.'

'Do they know whose carriage it was?'

'I'm afraid not. They didn't stop. Again, common, for if you stop you may have to pay. My officers have asked people in the area but have been unable to identify the driver. It was dark.'

She nodded. 'Tragic.' And yet also convenient.

'Indeed.'

Jeanne watched the hens picking at the grain. 'Her father must be bereft.'

'I imagine so, Marquise. I imagine so. But then people die every day.'

'And tell me also of the missing children saga. I hear Paris is in uproar.'

Berryer's face tensed, though his smile stayed in place. 'I would not say uproar, Marquise. There have been some disturbances, certainly – broken shop windows, a few fires, that sort of thing – but nothing that cannot be managed.'

'Nevertheless, it is worrying. How many children do they say have been taken?'

'Some are saying five, some say more, but of course most if not all of those will be runaways.'

'That's what the people say, is it?'

Still the smile. 'Oh, they say all sorts of nonsensical things; the rumours grow ever more ridiculous.'

'And what are the rumours?'

'Really, Marquise, they are too ludicrous to speak of.' He laughed.

'Nevertheless, you will speak of them.'

He abandoned the smile. 'Some people say that children are being murdered and their blood used as a cure.'

Jeanne felt her throat tighten. 'A cure for what, exactly?'

Berryer grimaced. 'Leprosy. Nonsensical, I know.'

'There is more?'

A pause. 'It is being claimed in some quarters that the

229

children are being brought to Versailles. That it is the nobles here who are in need of all this blood.'

Jeanne swallowed. She thought of the *poissonade* claiming she bought potions from witches; remembered how they'd accused Montespan of preserving her youth with blood. 'Does Louis know about this?'

'Of course not.'

'No. We won't tell him. It will only upset him. But these rumours must be squashed like flies, Nicolas. They must be exposed for the folklore they are. They cannot be allowed to grow more pervasive.'

'I agree, Marquise, but I do not see how I can stop such—'

'You can stop such stories, Nicolas, by finding a more likely theory for them to discuss.'

Berryer raised his eyes to hers. 'I do not think—'

'There are other rumours, as I understand it, that your officers themselves are behind the disappearances. This is unhelpful.'

'Deeply unhelpful, Marquise. And untrue.'

'Then you must find those culpable and make an example of them. The mood of public anger and uncertainty is dangerous. It does not make for a happy populace.'

'We do not know who the kidnappers are, if indeed they exist at all.'

'Nonetheless, you will find them and punish them.'

His tongue flicked over his thin lips. 'Yes.'

'Good.' She threw another handful of grain. 'Do you know, Nicolas, that I fancy it is time for me to change my tack at court.'

'Oh yes?'

'Yes. A new plan is needed, I think, to make me more secure in my position.' Which was currently highly precarious.

'I will of course do what I can to assist you, whatever course you decide.'

'Nicolas, you are too kind. I sense that things will turn out well for us.'

He bowed again, but kept his eyes on hers, watchful as a bird of prey.

19

Madeleine

Grief, it seemed, took Doctor Reinhart like a sickness. He didn't eat his meals. He didn't much sleep, nor speak. He retreated to his workshop where Madeleine could hear the occasional clatter of metalwork, the rasp of a saw. Often, however, there was simply silence. She imagined him sitting on his own, unmoving. Occasionally Lefèvre would arrive and insist on being admitted for a short time, now and again the cabinet-maker or the jeweller would visit, but mostly Reinhart was alone.

'He'll take ill,' Edme fretted. 'I make him the choicest morsels – all of his favourites – but he won't eat them, not a mouthful.'

The servants ate them instead: buttered lobsters and terrines of wild hare, caramel tarts and Hanover ham. It was a joyless kind of eating, though, Madeleine thought, the meat sticking to the roof of her mouth.

Distress and shock seemed to have drained Reinhart even of the energy to find out who had been driving the carriage that had run Véronique down. On one of the rare occasions that he appeared at the dinner table, Madeleine summoned the courage to ask again if he'd heard further from the police, for it had been a week now since Véronique's death.

'Surely, Monsieur, they must have found something.'

'They've discovered nothing, or at least so they say. Perhaps they are too busy keeping down the riots. But what does it all matter now in any event, Madeleine? Véronique is dead.'

'Do you not wish to know who it was? Not wish to punish them?'

He gave a short laugh. 'Punish.' He pronounced the word carefully, as though tasting it. 'What punishment do you think would be appropriate were we to find this person, Madeleine? What do you think we should do to them? Do you think it would bring her back?'

She didn't know what to say to that. It was as though the light inside him had flickered out, like a lamp run dry of oil, while she herself seemed to be returning to life at the very moment she wished she to be numb.

Without agreeing upon it in so many words, Madeleine and Joseph took it upon themselves to do the job they suspected the police had only done with half a heart, if indeed they'd bothered at all. Together, they went to the Place Baudoyer to find the spot where Véronique had died. It was indeed a busy corner, with carriages vying for space with the carts and cattle making their way to market. As with most places in Paris, there were no pavements, so pedestrians were forced to walk at the side of the roads, negotiating their way through mud and around piles of dung, flattening themselves against walls when a carriage came too close.

Madeleine tried a brandy-seller first: a man whose grog-blossomed nose advertised his taste for his own product. He frowned when they asked him if he recalled a girl being knocked down and killed just over a week ago. 'Killed, no. I heard of a girl hit by a horse not long ago. And there was a boy a month or so since, crippled under a wagon. Just over there.' He pointed. 'The man in the wagon – you think he apologised? You think he said sorry? No, he blames the boy. He blames the boy's mother. She should have been taking better care of him. He shouldn't even have been out at this time. *Tout ça.*' He waved his arm to indicate the ridiculousness of it.

'But not a girl?' Madeleine persisted. 'You haven't heard about a girl being killed?'

He pushed out his lips. 'No, I never heard that. But you know, I'm only here now and again. I walk the other squares. And the talk at the moment is all of the stolen children, not the accidents that take place on the streets. You should ask one of the shop vendors.'

The *tobacconiste* was no more helpful and refused to let Joseph enter the shop at all. He'd seen no girl being knocked down recently, but yes it was quite possible it had happened and he hadn't heard about it. He was a busy man. He didn't have time for tittle-tattle with other shop-sellers, nor indeed with them. 'This corner is well known for its accidents. We've told the Watch, time and time again, that they must do something about it. That they must erect bollards, or a gate. That they must find a way of stemming the tide of carriages through here. It's too narrow. But do they listen? Do they do anything? Of course not. They don't care if a few hundred are trampled every year. What is it all to them? They don't even bother to find the children who are being taken. The lives of the poor are cheap.'

'But no girl? You don't recall a girl being killed by a fiacre; by a carriage that kept on driving?'

The man shook his head. 'I never saw it, but I don't doubt that it happened. Ask Madame de Jontin in the *boulangerie*.

234

She's a good one for watching people, nosy old sow that she is.'

But they had no luck at the boulangerie either, nor with the local *bouquiniste*, selling books and pamphlets from his stall. People could tell them about others injured, about the growing unrest in Paris, and about fights and falls and frauds, but not about a girl killed here – not about Véronique. Madeleine and Joseph walked back together in silence, the light fading, the lamplighters climbing their ladders, people hurrying towards home. Another notice had been posted on the Porte Saint-Denis: 'People of Paris: do not let your children go out alone.'

'Can a girl really have died here and people not heard about it?' Madeleine wondered, as they continued towards the Louvre.

Joseph shook his head. 'I couldn't say. The master said the police told him this, but maybe they have told him the wrong place. Or maybe he misunderstood.'

'Perhaps,' Madeleine said.

But perhaps there'd been no carriage accident. Perhaps that's not how she'd died at all.

Back at the Louvre that night, Madeleine decided to go once again to Reinhart's workroom, to see if there was anything further to be found. Camille wouldn't let her off her task just because her mistress had snuffed it. And, more than that, she'd a creeping feeling that Reinhart had only told them part of what he knew, leaving the rest in darkness.

Just before midnight she padded barefoot down the stairs, her lamp trailing into the gloom. Once again, she tried the key in the lock, but this time, though the key turned, the door wouldn't open. She pushed at it, pushed again, but nothing gave. Testing the door, she felt that it was jamming further up and, raising her lamp, she saw another lock, one she hadn't noticed before.

Reinhart must have added a further lock or used a key he hadn't bothered with before. She retreated to the stairwell, her heart shuddering. The doctor must have worked out someone was rifling through his room. Did he know that it was her?

Madeleine stopped halfway up the stairs and looked back at the room, the doorframe just visible in the half-darkness. Or perhaps that wasn't it at all. Maybe there was something in that room that he wanted to hide, more closely than even his machine: something to do with Véronique and how she'd really died.

*

A month after Véronique's death, Doctor Reinhart emerged. He seemed to have been parched by loss, his skin taut over the angles of his face, his chest concave beneath his brocade waistcoat, and through his jet-black hair ran a new thick shock of silver. But there was a nervous energy to him that Madeleine hadn't seen for weeks. His eyes flickered behind his glasses, his hands twitched. 'It is done,' he told the servants that morning. 'My project for the King – it is complete.'

There was to be a great unveiling, he announced. A great show. Madeleine's throat tightened. A great presentation of the machine she had tried and failed to see.

'But what is it?' Edme asked nervously. 'What is it that you're unveiling, Doctor Reinhart?'

He smiled and put his finger to his long nose. 'It is a surprise. I cannot tell you or the surprise will be spoilt. And it is the surprise which is half of the purpose of my creation.'

Edme and Madeleine looked at each other, neither trusting his judgment. He seemed overexcited, feverish, his grief morphed into something strange.

'Will anyone else see it before it's revealed, Monsieur?' Madeleine asked lightly.

'Not a soul,' Doctor Reinhart answered. 'Not even Lefèvre has seen the adapted version, which of course has annoyed him terribly, given he thinks the whole enterprise was his idea and is anxious to claim credit. But I think, ultimately, he will be pleased. I think he will recognise my achievement. Now,' he clapped his hands. 'Let us have rolls and chocolate. I find I am suddenly famished.'

He came to find her later in the day, as she was cleaning the porcelain. She didn't notice him until he was a mere foot away. 'I will need someone to help me in case anything goes wrong,' he said. 'An assistant.'

She looked up at him. 'An assistant, Monsieur?'

'Yes. The arrangements are being made. I will need Joseph to remain here, so you will need to come with me. On Tuesday. To Versailles.'

Madeleine nearly dropped the plate she'd been holding so that it clattered on the table. Versailles, the sparkling palace that Maman and Coraline had discussed endlessly but to which she herself had never been; the centre of fashion and etiquette, a castle built of silver and gold.

'Certainly, Monsieur, if you're sure that's what you want. But if I'm to go to Versailles, then—' she gestured to her plain dress.

'Well, it is not a banquet, Madeleine. You will not be there to dance. But yes, you're right. They have rules. You should wear something fine. Plain but fine.' Reinhart was drumming his fingers on the wooden table. 'I will give you some money to obtain something suitable. No flounces, though. No bows. None of that nonsense. You will be on hand to help me with the mechanism, not be a part of the show itself.'

The worker of the mechanism. But what, *au nom du Christ, was* the mechanism? Was it truly a human, as Pompadour had suggested, or something even worse than that? She thought of the tray of coloured glass eyes she'd seen. The drawing of a cage.

'What will you need me to do, exactly, Doctor Reinhart?'

'I will need you to carry various pieces of equipment and run for assistance should anything go amiss with the machine.' He looked at her over his spectacles and wrinkled his nose. 'I warn you that it will be difficult to watch at first, but necessary. Know that I have my reasons for doing what I am doing.'

This made Madeleine very uncomfortable, nervy as a cat, so that the excitement of Versailles almost fled from her. For if it was something terrible and she'd failed to warn the police, Camille would finish her for good. 'Can you not tell me what is, sir? Just so that I'm properly prepared?'

He seemed to think about this for a while, then shook his head. 'In this city, Madeleine, all things, all people are connected. No words are safe, even if whispered, even if told to your closest friend. You will see it in two days' time, and so will all the world.'

Two days. She would visit Émile tomorrow. And she'd try to bring him back a souvenir from Versailles: an offcut of fabric, a sugar mouse, something to make him laugh. When Madeleine returned to her work, she saw that the plate she'd set down had cracked, a fine line running across the width of the porcelain like a scar. She went cold at that. She didn't like to think of herself as superstitious, nor given to fancy, but this seemed to her a bad omen. And more to the point, she'd broken a plate – Edme would have her skin.

Late that evening, as bats flitted across the charcoal sky, Madeleine decided to make one final attempt to get into the workroom. Using a stool that she'd carried from the kitchen into the courtyard, she climbed into the tree that grew beneath the workroom window. She'd have done better, however, to bring Émile, for she struggled to pull herself from branch to branch, and she cursed at her wretched skirts. More than once she nearly fell, scraping her shins, and by the time she pulled herself up onto the windowsill her stockings were torn and bloody and her nails were ripped to shreds.

As it turned out, it was all for nothing. The window was locked, just as she'd feared it would be. After resting for a moment she moved closer to the glass. She could see little at first, save for the twisted reflections of the branches and the dark buildings beyond. And then, suddenly, out of the darkness, the white oval of a face. She drew in her breath to scream, but in the moment before the sound left her body, she realised that it was not a real face she was looking at, but a mask. She exhaled, shakily. It must have been one of the masks Reinhart kept in his cupboard, brought out into view. But why? Perhaps it was nothing, but that feeling of unease stayed with her not just that night, but all the next day, as she counted down the hours until they would leave for Versailles; until the machine would finally be revealed.

Three

Versailles

20

Madeleine

The whole route to Versailles was a two-way stream of ornate carriages and shining horses, trotting past a flashing river and mostly empty fields. Now and again they saw peasants, filthy and stunted, hacking at the land, or turning to squint at the carriages. Mostly, though, the land was vacant and quiet, the only sound the hooves and the wheels.

After an hour or so, the traffic slowed, the horses snorted, and Madeleine's chest tightened with anticipation and fear. They must be nearly there. She, damaged daughter of a bird seller, was about to enter the greatest palace in all of Europe. Over two thousand rooms, Coraline had told her. A hundred staircases. More mirrors than you could count.

Then, as they climbed a hill, the palace reared before them like a drunkard's vision, the gilded tops of the roofs and cornices glinting gold in the sunshine. The sight was like a blow to the heart. From this distance it looked like some miraculous concoction of spun sugar and caramel; as if she could eat it all up.

'*Mon dieu*,' she breathed and Reinhart, who'd remained silent for most of the journey, turned his eyes towards her.

'It is a symbol of royal power. It was built to impress, to intimidate, to remind the King's subjects of their lowness and inconsequence, their own meagre mortality.'

Well, it did that all right. Madeleine felt a stab of anxiety. Wouldn't they recognise her for what she was in there: false paste in a palace of gems?

But then Reinhart said, 'You will see that inside it is rather wanting. More to the point, it stinks.'

As the carriage reached an immense oval-shaped courtyard, Madeleine saw, still far off, the main palace: a giant's castle, the delicious colour of honey. A long walk led to a fountain on a wide terrace, and beyond that an ornamental lake glimmered silver in the morning sun. Everything was built on a vast scale, so that the people walking about it looked like so many ants, crawling over the land, insignificant and ridiculous on this monstrous display of wealth. Their coach was met by dragoons, who escorted them to a barrier where their papers were checked, and then into another courtyard, and another stopping post, until at last the carriage door was opened and Madeleine, journey-rumpled, descended to the ground.

Uniformed men came to help unload the machine, which was in two brocade boxes and, by the looks of the men trying to lift them, uncommonly heavy. They passed through another gate and entered Versailles itself. It was as Madeleine imagined it would feel to enter a foreign country, or perhaps another world. All about them, courtiers walked among the orange trees, women arm in arm, their steps so small and so light it was as if they glided on unseen wheels. Their panniered skirts were as wide as a fiacre and their pastel powdered hair piled so high that they seemed almost a head taller than the people of Paris. They spoke differently too. The lines of conversation Madeleine snatched from the air seemed alien, made up of words she'd never heard. And the laughter, everywhere ripples of laughter, high and clear and false.

As they walked, she glimpsed glittering salons, golden galleries, gilded paintings; even a golden kennel complete with a burnished hound. And, everywhere, people with lace at their cuffs and jewels at their throats strutting and jostling and preening. In one courtyard a finely dressed man – a prince, for all Madeleine knew – sat before a portrait painter, very still and very regal, while courtiers stood by to watch. In another, a group of women were practising at dancing in dresses of pure green silk.

It all looked beautiful, but, like Reinhart said, it stank. Whether it was the drains or the river, or the people themselves, Madeleine wasn't sure, but there was a rank smell about the palace, a note of putrefaction beneath the perfumes and flowers. There was something else odd about the place too, though Madeleine couldn't quite put her finger on it – something important was missing.

Doctor Reinhart seemed to know where he was going, leading the men who pulled the wheeled boxes. She followed him through another courtyard lined with lemon trees in silver tubs, and then up a stone staircase that reeked of piss. At the top of the steps two men in livery were waiting, and took them through to a gallery where men and women milled about drinking wine and reclining on chairs eating marzipan fruit and candied rose-petals from gold and silver boxes. A miniature goat with a ribbon at its throat gnawed on the leg of a chair. In a corner, she saw a woman sitting back in an armchair, her great skirts puffed out before her, her eyes closed, groaning, apparently in pain. It was only after a moment that Madeleine noticed a man's red-heeled shoes poking out from beneath the skirts and realised that the woman's groans were not pain at all. Then on again along a corridor and into an astonishing room of mirrors and lights, which Madeleine passed through as though in a dream.

They came at last to their destination, two men in blue livery opening the doors to reveal a high-ceilinged salon painted white and gold, with murals of gods and goddesses above the doors. A huge chandelier dripped crystal from the ceiling; crimson and gold hangings lined the walls. The room was empty except for a striking young woman in mint silk with a parrot on her shoulder, like some kind of royal pirate. 'I wondered when you'd come,' she said. 'They said you were putting on a show.'

Hearing a squawk from above, Madeleine looked up to see a scarlet macaw perched on a picture frame. 'I let them fly in here,' she said haughtily, seeing Madeleine staring at them. 'I can't let them go outside.' Then to Doctor Reinhart. 'What is it, then, that you're showing? What is in the boxes?'

Doctor Reinhart frowned at the girl. 'I'm afraid I cannot tell you, Mademoiselle. It is a surprise, for the King.'

'But I'm his daughter.'

He smiled. 'All the more reason not to, your Highness.' He bowed.

The girl frowned, stood up and whistled and the macaw, after a moment, flew to her. She returned the birds to a golden cage and closed the cage door, then walked from the room, looking back once at the black cases, glancing suspiciously at Reinhart.

He rubbed his hands together. 'Madeleine, you may leave me for a few moments. To prepare.'

'Can I not assist?'

'No, this is something only I can do. Come back at half past the hour.'

Madeleine took the steps back down to the courtyard with the lemon trees, where a group of men were sitting at a table, playing at dice. They eyed her idly and continued with their game. Not knowing what else to do, she sat by a fountain, watching the golden fish darting beneath the surface, listening to the trill of the water and to the conversation of the players, the fall of the dice. Now and then men and women in twos or threes would wander through the courtyard, on their way, perhaps, to the salon. As Madeleine stared at the statue of a nymph in the centre of the fountain, she realised with a chill what was wrong about Versailles, what it was that was missing: among all the people parading about, there wasn't a single child.

As the minutes went by, the number of people filtering through towards the salon grew, and the fluttering in her chest grew stronger. She saw valets in blue livery walk briskly across the corridor holding silver platters of food, silver pitchers of drinks, and, a little later, a quartet passed by, carrying their polished instruments. Her heart began to beat faster as she knew that her hour, and that of the machine, approached.

At half past three she duly returned and could hear the murmur of conversation from beyond the door. The room was now filled with people, men and women in exquisite concoctions of silk and lace, with jewels in buttonholes, on buckles, in hair, or resting or nestled in the many bosoms on show. The smells of expensive scents mingled with the rich odours of the food, and with the rank smell that pervaded the palace. On buffet tables, food had been laid out: roast chickens and caramelised ducks, pyramids of candied fruit, piles of patisseries and cakes. Waiters circulated with trays of wine glasses, decanters of liqueurs, silver coffee pots. In a corner, a small dog lifted its leg and pissed against a marble nymph.

At the front of the room, Doctor Reinhart stood with his hands clasped behind his back. Beside him stood what must be the precious machine. It was immense, a good head taller than him, shrouded in a velvet cloth, beneath which Madeleine could see a large pedestal. She was relieved to see Lefèvre standing near Reinhart, offering her a reassuring smile. Reinhart only nodded when he saw her. A sheen of sweat had broken out across his brow. It wasn't only she who was nervous.

'If I motion to you, you must come and do exactly as I tell you. You understand?'

'Yes, Monsieur. I understand.' Her mouth was dry as dust.

After a few minutes, a hush spread across the room and then the crowd parted. It was the King, a blue satin sash across his jacket, a smile upon his face. Beside him walked Madame de Pompadour, a vision in pink and rose, diamonds at her neck and in her hair, her eyes surveying everything. Louis took a gilded seat at the front of the room, and Pompadour a seat beside him. He raised his eyebrows at Doctor Reinhart and gave a slight smile.

'It is ready?'

'Yes, Majesté. It is ready.'

21

Jeanne

Whatever this infernal machine was, it surely couldn't be any worse than the things Jeanne had imagined, though who knew what the clockmaker was capable of. He was certainly an odd-looking man, she thought as she regarded him: tall and cadaverous with an unhealthy gloss to his face. At first she thought him ill, but perhaps he was merely nervous. She knew how nerves could make one sick.

'Ladies and gentlemen,' he said, his accent Swiss, 'I present here for the first time a creation I have been working on for some months, with the kind patronage of His Majesty the King. A creation which is the first of its kind and which I humbly suggest goes further even than the automata of Jacques Vaucanson, for it is not just the outward organs of this auto-maton that move. Vaucanson created for you a flute player. I give you ... The Messenger.'

As he spoke, he pulled back the cloth that covered the machine, and, from the crowd, there came a vast intake of breath. There before them, raised on a pedestal, was a beautiful girl in a pale silver dress, a golden crown glinting on her head. She sat at a small, gilded writing table and in her hand she held a quill. The lights from the many candles danced upon her porcelain face and it was almost as though she breathed. No, dear god, she *did* breathe, or at least her chest moved – a barely perceptible motion that made Jeanne's blood run cold.

After a few moments, the clockmaker climbed onto a stool behind the automaton, inserted a golden key and turned it several times. There was a clack, a whirring noise, a long moment of silence, and then the doll began, sickeningly, to move. First, she tilted her head to one side, as though considering. Then she lifted her hand, dipped her pen in a bottle of ink, and put the nib to the paper before her. Murmurs from the crowd; astonished laughter. Then came the scratch of the quill as she began to write. And as she wrote, her eyes slid from one side to the other, as though following the words. It was like the horrible doll that Jeanne had disposed of, only the mechanism perfected, the eyes a glassy green. After a few moments the automaton paused, tilted her head to the other side, then wrote again. Finally, she lay down the quill and looked up, at her audience.

At this sign, Doctor Reinhart walked forward and picked up the piece of paper. He held it up before the crowd then passed it to Louis. Louis stood to read the paper. 'She says:

'I may be made of gears and cogs,
I may seem false to you,
But I'm as alive as any girl
And what I write is true.'

A cheer went up, thin at first, growing richer and more confident as a smile broke out on Louis' face and he began to laugh. Jeanne felt a terrible nausea sweep through her, felt the acid rise in her throat. For Louis must see what she saw, must have known about it in advance: the doll, with its perfect porcelain face, its long fair hair, its silver dress, it looked just like the girl she had seen in the clockmaker's shop. It was Véronique, the clockmaker's daughter, raised up from the dead.

In the minutes that followed, as her heart rate slowed, Jeanne had time to take closer account of what she saw. It was Véronique and yet it was not Véronique. It was her face, painted in white, red and pink to recall her flawless complexion, her youthful flush. They were her eyes, the same fire-centred green that she had envied when she met her, but crafted from glass, shining blankly.

She watched Doctor Reinhart as he rotated the stand and lifted the doll's dress, which had been cut in half, so that the audience might view the empty cavity at her back. She watched as he reached into it and showed them the mechanism by which he had programmed her to write: a lattice of rods, wire threads, and in the centre, a column of small brass discs. Jeanne saw that Reinhart's maid – her little spy – stood not far from him, staring at the doll. She had not noticed her previously, so intent had she been on the machine. The maid was dressed far more finely than when Jeanne had seen her at the clockmaker's shop, in a linen dress with white lace at the cuffs, a veil in front of her face, but even so Jeanne could see that the girl was pale with fear.

'I can change the letters,' Reinhart explained to the crowd. 'She can be made to write anything that you wish.' He smiled. 'It is, I believe, the first time it has ever been done.'

When Jeanne spoke, her voice came out too shrill, too tight. 'Then have her write a message for our King, for our Louis.' She stood and walked forward to Doctor Reinhart, leaning in to whisper to him. He smelt of musk and powder and something medicinal. Reinhart frowned, then nodded. He turned the doll back to her previous position so that he might work the clockwork without others seeing. He passed a hand across his brow. Again, his lips worked as he talked to himself, perhaps reminding himself how to operate his creation, or perhaps talking to the doll. After what seemed like an eternity he seemed satisfied and climbed up again to wind the key.

A pause, and then, as before, the doll tilted her head, lifted the pen and began to write. Once more the room fell to a hush as the pen scraped across the paper, only the whisper of the nib on parchment. She put down the pen, tilted her head to the other side, and in the light almost seemed to smile.

Reinhart nodded to Jeanne and she walked forward to collect the paper. As she read what was written, her mouth moved into a wintry smile and then she tipped her head back and laughed. 'Our new friend is very clever, I think.' Too clever. 'She says:

'Forgive my lack of etiquette
For I was not properly taught,
I beg the King's permission
To be introduced at Court.'

Laughter from the audience. Then clapping. Jeanne passed the paper to Louis, who read it himself, a look of annoyance upon his face.

'Does she laugh at me, your writing doll?' he asked. An ego so fragile his pride could be hurt by a porcelain and metal doll.

Doctor Reinhart shook his head. 'Oh no, your Majesty. Quite the opposite. She weeps.'

Jeanne heard Louis gasp. For it was true; she saw it before the rest of the audience did. Tears had formed in the corner of the doll's eyes and now rolled down her painted cheeks.

'*Mordieu,*' Louis said, leaning forward. 'So she does. You have done it. A doll that not only writes and breathes, but cries!'

She waited until the audience had drunk their fill; until they had wandered close to the doll, and murmured and gazed up at its face, and peered through the cavity at the mechanism. She waited until Louis himself had taken Doctor Reinhart by the shoulder and told him he was quite the genius and how on earth had he done it in so little time? Doctor Reinhart did not answer his questions, merely smiled, the grin like a rictus on his bone-pale face.

A few members of the crowd hung back, staring, uneasy, but only two obviously shared her revulsion: her maid, who stood more rigid than the now quiet doll, and the King's own tutor, the surgeon Lefèvre who, once the performance was over, left the room hastily, eyes down, his complexion drained to ash. No doubt they were wondering the same thing that Jeanne was: why would the clockmaker do such a thing? Why resurrect his own daughter?

Louis gestured to one of his guardsmen, a man in scarlet livery, very tall and thin, who bent to listen to what the King asked him. The King gestured to Doctor Reinhart, to the doll. The man bowed and left. To Reinhart, the King said, 'I want everyone to see this. She is to stay here, in my clock room. I want you to show her again, starting tomorrow.'

Reinhart nodded. 'Certainly. But she needs a place of her own, where she can be kept safe. Where no one can touch her.' He ran his fingers over her waist.

'Very well. That can be arranged. We will bring people to move her.'

'I will accompany them,' Reinhart said. 'Madeleine here will assist. The mechanism is very fragile. She must be handled carefully. No one must interfere with the pedestal.'

'Yes, of course.' Louis touched the doll's hand thoughtfully. He smiled. 'My word, but it feels almost like skin.'

Reinhart blinked. 'Yes.' Then, very gently, he folded the doll's hands, bowed her head, and threw the veil over her so that, once again, she was blind.

★

'Was it your idea, *mon cher*?' Jeanne asked lightly that evening, when they were taking supper together in the *petits apartments*. Brillant, Louis' white angora cat, sat on the chair beside them, eating the pieces of meat the King occasionally tossed her way.

'In part, yes. We had discussions, you see, about what had been achieved so far, what was possible, what would advance science, what would further understanding of how the body worked. When I realised how skilled Reinhart was, I tasked him with creating an automaton that went further: that not only moved externally as Vaucanson's had, but breathed and cried and bled.' He threw another bone to the cat. 'Of course the bleeding he doesn't seem to have managed, but arguably he has gone further than that, as the doll appears to think. I know some of the ignorant herd will say we have gone too far; that we are playing at Gods, creating humans like some latter-day Prometheus. That is why I kept the enterprise covert.'

'No, *mon coeur*, I meant: was it your idea that it should be *her*?'

Louis frowned and picked up another chestnut-coated chicken wing. 'Ah, I see. No, I merely suggested that it should be a girl. To make it Véronique was Reinhart's own notion, I think. After his daughter died he said he wanted to dedicate the automaton to her, so I suppose he adapted it to look like her. A tribute, if you like.'

'Do you not think that a little peculiar, my love? That a father should wish to recreate his lost daughter?'

Louis shrugged. 'Not really. Descartes was said to have made an automaton of his little daughter, Francine.'

'I always thought that a fable.'

'Perhaps. But many people recreate a version of their departed loved one. Portraits, keepsakes. You forget that I myself had a bust made of Pauline.'

Oh, but she had not forgotten. The bust of his former mistress still graced the mantelpiece in the *petits apartments*, though she had often longed to smash it to the ground.

'It does well to remind ourselves of death,' Louis continued in an indulgent tone. 'Of the value of life. The Queen gazes on a *belle mignonne* every day to impress upon her mind the vanity of worldly things.'

Jeanne shuddered inwardly. 'Yes, of course.' *Belle mignonne*, indeed. It was a ghastly skull, ornamented with ribbons and dripping candles. The woman grew almost as bad as her husband. 'And yet, there is something, I don't know, slightly distasteful about the doll. I note even your surgeon looked somewhat taken aback.'

'Oh, I suspect Claude is merely annoyed that it is not he who invented the doll, and that he had only a minor part in it. He has some notion it was his idea, you see, though really it was mine. And he likes to be seen to be at the forefront of things. To be regarded as a magnificent mind.'

'Perhaps. Or perhaps he thinks the idea distasteful. Why is she dressed like a princess?'

Louis frowned again, licking his fingers. 'It sounds almost, Jeanne, as though you were jealous of a girl made of metal and gold. Why must you always spoil my fun? Why must you cast doubt on the sole thing that has cheered me, the one venture that has diverted me these past weeks?'

A cold snake of fear writhed in her stomach. 'I do not wish to spoil it, I am only—'

'I know you too well, Jeanne. Just because you have had no part in this project, you must throw scorn on it and insist on finding some other elaborate game for me. Well, I am quite content with this one for the time being. Far more content, in fact, than I find I am with my current mistress.'

Jeanne felt icy dread slither from her gut to her chest. 'I'm sorry, *mon bijou*. It is not that I doubt its ingenuity. I am just trying to protect you.'

He put out a hand, still greasy from the chicken. 'I do not wish to hear your opinions on this. It is not something I should have expected a woman to appreciate, certainly one without any scientific learning.'

Jeanne swallowed down the words that were forming: that she had in fact read widely on anatomy and mechanics, that her own private library was stocked with Morocco gilt leather books on those very subjects. That, if anything, she knew more about them than he did. But she did not say any of this. Louis' opinions were never to be questioned outright.

He turned away from her. 'Let us talk no more of it, else you will drive me to say something we will both regret. Please try to be less tiresome.'

Jeanne felt a lump in her throat, formed of rage and fear. She had thought that with Véronique gone, she would be safe at last. But, as in a nightmare, the girl had returned, more powerful than ever.

22

Madeleine

Madeleine could barely talk to Doctor Reinhart now, scarce look at him. All that time he'd been hidden away, all that time she'd imagined him grieving, he'd been creating this 'thing' — this perversion, this sham — of his clever, life-filled daughter. And because she'd been unable to guess at his scheme, Camille would claim that she'd failed.

It was a mercy that she'd been housed far away from the man, assigned a tiny room high up in the rafters of the palace, with space only for a bed, a commode, a basin for washing, and all to share with another maid. Up here, away from the perfumes and beeswax, the potted orange trees and vases of flowers, the true smell of Versailles won out: stale urine, stale bodies, tallow candles and shit. The floors here were filthy, unscrubbed flagstones and uneven wood, and the linen a dirty grey. It struck Madeleine that the grander the residence, the narrower the domestics' quarters, as though the richer the masters, the cheaper they valued the lives of those that served them.

There was the view, though, like a window onto an enchanted land. A great park, a series of lakes, and behind that, the silver waters of the Grand Canal. She slept poorly that first night, partly because the other maid cried out in her sleep, mainly because she thought again and again of the mechanical doll, its face too like Véronique's, its movements uncannily real; its eyes that seemed to stare at her, but at the same time to see nothing. As she lay in the darkness, she tried to work out when Reinhart had started to create the doll, and when it had morphed into being his daughter. Had Reinhart only adapted it after she'd died, or had he always intended it to be her? If he had, he'd kept it to himself, for she'd seen Lefèvre's horror when the doll was unveiled – he couldn't leave the room fast enough. The King, though, the King hadn't blanched; in fact, he'd seemed entranced. She thought of little white Franz and the rabbit that came before him, of the different geese Reinhart had procured until he found the one he wanted. She thought of Émile lying alone in his bed and wanted desperately to go to him. When she dreamt, it was of her mistress, running through midnight streets, then of Suzette with a porcelain face.

She woke early in the morning to the creaking cry of the waterbirds. Through the open window she could just about smell the lilac trees above the piss. Then began the sounds of the palace as it began to stir itself from sleep: the distant clatter of iron grates, the tramp of military boots on the gravel outside. A short time later came the ringing of bells and a series of shouts, summoning servants to their various tasks. There were no bells for Madeleine, though, and no one sent for her. A gnawing having set up in her belly, she decided to make her way down to the kitchens to find herself some breakfast. This, however, was no easy matter, for Versailles was like a giant warren, of narrow corridors and servant's passages, some well lit, some dark as the devil's arsehole. The other servants all seemed to know what they were about, running here and there carrying barrels of water and plates of food, jostling her, swearing at her for getting in their way.

At last she found the kitchens, where a great fire roared, and where tens of red-faced cooks were at work stirring huge copper pots and shouting at scurrying underlings. It was, she thought, like a vision of hell, and very nearly as hot. She was hounded from the kitchen, shown the place where the servants must take their food and where thirty or so men and women were seated at long tables, chewing, chattering, squabbling. As she sat eating her brioche, a solid-looking woman approached her, well dressed in dark blue silk with pure white lace at her throat. Not a regular servant.

'You're the clockmaker's maid,' she said quietly.

'Yes, Madame.' Madeleine stood up.

The woman shook her head. 'Sit back down. Finish your breakfast. But when you're done, you're to come to the Parterre d'Eau, the side closest to the palace. I'll be waiting for you.'

Madeleine had no notion of where the Parterre d'Eau might be and the people she asked were too busy or too rude to answer her question, or had little idea, being entirely lost themselves. When she finally found the place, the woman in dark silk began to walk in step with her. 'I'm Madame du Hausset,' she said, taking her arm. 'I was told to fetch you. We will pretend we are great friends taking a promenade in the gardens.' She began to walk her swiftly around the marble-rimmed lake, past reclining nymphs and stone-frozen children.

'But where are we going, Madame?'

'We're going to the Labyrinth. I'm taking you to my mistress.'

'The Labyrinth?' The whole of Versailles seemed a bloody labyrinth, its corridors and rules a riddle. On some doors you must knock, on others you must scratch; in some rooms you might sit, in others only stand, no matter how your back might ache.

'Don't worry,' the woman said. 'It's not so very far.'

What could Madeleine do but go along with it? The gardens were astonishing: exquisite, impossible, endless. They hurried past what looked to her to be an immense waterfall, cascading over stone stairs, the sound so loud in Madeleine's ears she thought they might burst, then skirted an orangery where the air was at once filled with fruits and more jets of water, enough to bathe half of lice-bitten Paris, more water than she'd ever seen. Behind the water she could hear the notes of a song – the strains of a string quartet.

The Labyrinth was, as it turned out, just a maze of green privet, precisely clipped into lines. Du Hausset fairly marched her along the long rows, past gilded statues of Aesop and Cupid and a whole herd of animals, until they came to a fountain covered in birds. They weren't real birds, she realised, but figurines, beautifully painted: swallows, swifts, sparrows, robins, surrounding a fierce-looking owl. From behind a line of hedge walked a woman in lilac carrying a lilac parasol and a very small dog. It was Madame de Pompadour.

'Thank you, Hausset. You can leave us now.'

The woman nodded and glided back along the path.

Pompadour put down the dog, which whimpered. 'Well, my little *mouche*. Here we are.' She smiled, but her eyes did not. They shone with a strange lustre. 'You didn't know it would be her, did you?'

'No. No, I didn't.' For an instant she felt again the shock, the sickness she'd felt at seeing her: Véronique but as a grotesque doll.

'Why do you think Doctor Reinhart might have done that? It's not quite a normal thing to do, is it now?'

Madeleine hesitated. 'I suppose grief can do odd things to a person.'

'Oh, I know that well. But it strikes me as particularly strange given how recently she died. What was his relationship with his daughter?'

'I'm not quite sure what you mean, Madame.' And she wasn't sure how much she should say.

'Well, were they close? She worked with him, didn't she? He trained her?'

'He did, yes.'

'But?'

But why did she want to know? 'I'm not sure he's a man you can get that close to. Véronique had been apart from him most of her life, in a convent.'

'So I understand.' Pompadour paused. 'Was he very distressed when she died?'

'Well, he seemed to be. Mostly shut himself up.'

'"Seemed to be". You mean you're not sure whether he was?'

Madeleine chewed her bottom lip. How could she say what she'd begun to think, the cesspit of her imaginings?

And then, as though the woman had lifted the lid off her skull and looked inside, Pompadour said, 'How much do you know about how your mistress died?'

Madeleine swallowed. 'Not that much: that she was knocked down by a carriage. That they haven't found the driver.'

'What else?'

'I don't know anything else.'

'Ah, but you think something else. You're just unsure whether to tell me because you think it dangerous.'

The woman, Madeleine thought, was faster than a card sharp. 'Well, I feared at the beginning that she might have been taken by whoever it is who's taking the children in Paris.'

'But you don't think that now.'

'I don't know what to think.' She wished, in fact, that someone would tell her, because the things she thought were wrong.

'Who told you of Véronique's death, little *mouche*? Where did this story come from?'

Story. Was it just a story? 'From Doctor Reinhart, Madame. It was him that went to identify the body.'

'The police notified him, did they?'

'I supposed it was the police. He had a note telling him to come to the *Basse-Geôle*.'

'This note – did you see it?'

'No.' She hadn't even thought to ask. What a fool. So shocked had she been, so upset, that the old Madeleine – the noticing sort – had vanished into thin air.

Pompadour nodded, but her eyes had narrowed. 'All such a terrible business,' she said. 'All such a tragic loss.' She smiled curiously at Madeleine and tipped her head to one side. 'What caused the injury to your face, Madeleine? Or, rather I should say, who? A customer?'

Madeleine felt heat rush to her cheeks. Pompadour knew, then. She, the greatest courtesan of all, knew what Madeleine had been, what she was. Everything. Or at least, nearly everything. 'Yes, Madame. A long time ago.'

The Marquise nodded slowly. 'Men. They know where to hurt us, don't they? And now you'd better hurry back to your master. They are showing the doll again at one.'

★

The second showing took place in the same room as the first, with its tapestries, silver mirrors, heavy portraits of haughty, velvet-clad royals. Versailles both fascinated and revolted

262

Madeleine: its gilded rooms, glittering jewels, powdered foot-men and painted ladies, all jostling together only a few miles away from where people died of hunger. She felt similarly about the doll, for her eyes were drawn to it again and again, yet whenever she looked there came the graze of fear, the wish that it was gone. From where she stood, Madeleine could see the doll only from behind, and it could very well have been her mistress sitting there, poised and ready to write. She couldn't bear to be near it, to see it breathe, see it move, while Véronique herself was frozen in time.

She felt all at once the pressure of someone's gaze upon her. She turned her head. Seated next to a long-noised man with a white rolled wig was Madame de Pompadour. Their eyes met and for a moment Madeleine felt a surge of understanding between them: they would bear it, they would pretend, but they both knew what they were about to see was not only strange, but grotesque.

The rest of the audience had no such qualms. The second showing went smoothly and much like the first: all those simpering aristos delighted and so charmed and oh how sur-prised, the King complacent and smiling, as he watched the reaction of his guests. The first message the doll wrote meant little to Madeleine but seemed to please the crowd:

I think, therefore I am.

'She is quite the *philosophe*,' said the man seated next to Pompadour, 'and it certainly makes you wonder. If a machine can be made to act as a human, can be made as clever – nay, cleverer – than a human, then what is there to stop the machines?'

The King didn't like that. He frowned at the man. 'Make it as clever as you like, Monsieur Voltaire, a machine still has no soul, and no engineering can make it otherwise. To suggest they might become dangerous is frivolous. We will always be able to control them.'

A ripple of murmurs and the man gave an unconvincing bow. 'No doubt, Majesté. No doubt. I defer to your greater understanding.'

As if to prove its cleverness, the second message the doll wrote was a riddle.

I am neither dead nor alive. Who am I?

'A ghost!' screamed a woman with violet-powdered hair. Laughter and ghostly whoo-ing sounds from the wine-flushed crowd.

'My wife between the sheets!' came a shout from a man at the back.

More laughter, clapping, but Madeleine was now doubly nervous. For 'neither dead nor alive' was exactly what the doll was. It was life and yet it was *not* life, and it was that quality that made her shiver, that made her wake in the night, afraid.

The crowd of courtiers didn't share her concern. They were chattering to each other, fluttering fans, sipping from crystal glasses, enjoying the entertainment. One man, however, stood very still. *Dear god*, it was Camille. Madeleine dug her finger-nails into her palms, felt her heart begin to race. He must've heard about Reinhart's creation and come to judge it for himself. Would he give some sign that he knew her? But he didn't seem to have noticed her at all – he was staring at the doll, unsmiling.

He went up to them after the performance, as she'd known he would. He waited until the other guests had ooh-ed and aah-ed at the machine, peered at it through their gold-handled lorgnettes and asked their dull questions, then he approached the doll and touched her hand, just as the King had done. 'Well. This is quite something.' Camille smiled at Reinhart, then looked at Madeleine from the corner of his peculiar eyes.

'Tell me, Doctor – how did you make it so realistic? How did you make it look so much like the real thing?'

'I have studied for many years, Monsieur,' Reinhart replied, 'in order to create automata that resemble real creatures.'

Camille rubbed his jaw, still looking at the doll. 'Yes, yes, but a human, a very specific human, that's something else entirely, isn't it, Doctor?'

Reinhart's jaw tensed. 'Jacques de Vaucanson made the flute player some years ago. This takes it further. It breathes, cries, allows me to direct its movements.'

'I never saw the flute player. Did you, Mademoiselle?' He turned to Madeleine.

She shook her head, willing her cheeks not to colour. 'No, Monsieur. I'm only a maid.'

Camille looked back at the doll and touched the fabric of her dress. Perhaps he'd realised what she had: it was Véronique's dress. Her finest.

Another man came forward, smiling, to ask questions of Doctor Reinhart. Camille stayed close to Madeleine and to the doll. 'What d'you think of this, then, little Miss Servant Girl?' His voice was low. 'And how could you have failed to work out that *this* was what he was making?'

'Because he kept it entirely hidden. I tried my best, I swear to you, but he added new locks to the door, and locked the windows too. There was no way to get in and I couldn't have guessed that he'd do this.'

He looked at her closely. 'You look done in. Are you sick?'

'I'm quite well.'

'No. You think what I think. You think: what kind of man makes a moving model of his own daughter, and what else would that man do?' He looked again at the doll. 'The hair. It's exactly the same colour as hers, isn't it?'

Madeleine nodded, shivering.

He reached up and touched it, rubbing it between his fingers, and as he did so, she had a very clear vision of the rabbit automaton, dressed in Franz's pure white fur.

Camille looked at her in the eye. 'I wonder,' he said. 'I wonder.'

That night Madeleine dreamt she was Véronique's golden rope-dancer automaton, the little girl jumping up and down on her perch. In the dream, she flew free of her box, just like Pompadour's automaton bird. She walked, then ran through a maze of corridors, knowing someone was following her, chasing her, and as she ran she changed from a mechanical doll into a real girl – a girl with long brown hair and worn leather boots, racing through the mysterious passages of Versailles. As she turned, the girl changed again and she realised she was not herself any longer, but the young street girl from the Rue Thévenot, thin, terrified, running through not the passages of Versailles but the narrow streets of the Paris slums, tall houses looming above, hooves thundering behind, the pounding of blood in her ears.

Madeleine woke in the early dawn, her head pounding, her tiny room swathed in grey. She thought of the sketches of a cage that she'd seen all those months ago, the rows of coloured glass eyes. She thought of the strange material she'd seen drying, and the whitewashed walls of the room seemed to pulse. What if it was Reinhart who had taken the children of Paris, in order to make the doll? What if he'd used the children, just as he'd used the rabbits, the hens, the bats, to design and build the machine, then discarded them, just as he had the animals? That would be why the children had varied each time, in age, in colour, in sex, as he'd tried to find the right template, just as he'd gone through two rabbits, three geese.

And then, though she tried to prevent her mind from reaching into the very depths of the darkness – she thought of Véronique, laid out so carefully in her coffin. Could he have decided that the correct template was his very own daughter – that *this* was how to make her perfect? Had he killed her in order to bring her back as a breathing, moving doll?

23

Jeanne

Jeanne wanted the doll destroyed and the clockmaker banished from Versailles. But there was no question of that — the automaton was Louis' toy, and Reinhart his peculiar pet. The doll had proved so popular that Louis had decided to make the next showing a state occasion in the Galerie des Glaces, with the Queen, the Dauphin, Dauphine and Louis' wretched daughters in attendance. Of all of those, she knew only the Queen would bother to be polite. She always had been. But then after bearing the man twelve children (four dead, four sent away), she was evidently grateful Jeanne had taken those duties off her hands, hands which she now used only for prayer and for playing dull games of Cavagnole. Jeanne knew what the others called her: 'Maman Putain' or, in lighter moments, 'Pompom'.

Inside the *Galerie* the press was extreme, the long mirrored room filled with fluttering fans, rustling skirts, wilting lace, and wax fumes from the thousands of candles that dripped from the myriad of chandeliers, sending gouts of wax down to the rugs below. Sometimes she was staggered by the beauty of the place — the light reflecting off the mirrors, the glassware, the silver, the naked skin, but today it was too much, like being trapped inside a giant lantern, held too close to the flame. Her temples throbbed and her eyes watered as she made her way to her seat, wanting to get it over with, to return to the quiet of her apartments. But of course it could not be that easy, because Versailles never was. Richelieu stood in her path talking to the minister d'Argenson, a man Richelieu hated only a little less than he did her. The men turned as she approached and glanced at each other in such a way that she knew they'd been talking about her.

'And what do you think of Louis' clockwork princess, Madame de Pompadour? We were just discussing it, the Comte and I. Unnatural, wouldn't you say?'

Oh yes, she thought it uncanny, peculiar, perhaps a portent of evil. She hated Louis' involvement in the whole enterprise and she wanted it gone. They were not, however, talking about the doll. 'I think it remarkable, Monsieur.'

'Really?' Richelieu shrugged his shoulders. 'I do not think it so very special. She will suffice to entertain the King for a while, but truly she is a gaudy piece, merely for show. Nothing in her head, no real heart beating in her breast. Merely a soulless machine.' He smiled. 'I would say the King will tire of her fairly soon.'

She gave him an acid smile. 'I think you have underestimated the doll, Monsieur. I'll hazard she is much cleverer that you have guessed. I think she will last longer than any of you. That she can learn new tricks and play new games while others around the King crumble into dust.' Jeanne drew herself up and walked smoothly away from him, towards her seat. Catching sight of herself in one of the many mirrors, she noticed the veins standing out from her neck, thin blue worms beneath the white. She would not outlast them all, that she knew, but it would not be Richelieu who finished her.

*

The King's daughters, Victoire and Adélaïde, were already seated at the front of the semi-circle together with the Dauphin and his baby-swollen wife. The Queen, however, was nowhere to be seen, and her chair had not been set out. It was possible, of course, that she was ill, but it occurred to Jeanne that it might well be something else; Marie Leszcynska was nothing if not religious, her every day built around prayers. For her, as for the Church, the doll might not be merely unnatural, it might be a crime against God.

The princesses turned briefly as Jeanne took her seat behind them but did not acknowledge her. After all this time their rudeness should have ceased to bother Jeanne, but given she'd only offered them kindness (pitying them the sad, cocooned lives they had lived, like fat moths), she could not help but feel the sting of anger. She was relieved to see Berryer arrive, flanked by two junior officers. He came to greet her, then took a place standing not far to her left. The heat by now was stifling, the smell of florid perfumes and crushed flowers mixing with sweat, wig powder and wax. The clockmaker seemed to share her distaste for the scene; perhaps the crowds alarmed him. He hovered near the shrouded doll at the front of the room, twisting his hands together, talking to Lefèvre, the rumpled-looking surgeon who'd initially seemed so alarmed. Reinhart's maid stood not far behind, observing the crowd, her scar visible beneath the veil. It was a shame; the girl had a good figure and a fine face, but someone had branded her as one might a horse.

At last, Louis arrived, his satin sash a shimmer of blue across his jacket, his smile for her so radiant that for a moment her heart lifted and she remembered the joy she'd felt when first she became his, in the days when she'd shivered at his touch, her body blossoming like a flower. He took the golden seat at the front between the princesses and at once began to tell them how surprised and delighted they would be. Jeanne looked away, not wanting to hear their fawning responses and obedient laughter. 'I believe, Adélaïde, that the doll has a message specifically for you today.'

'Surely not. How exciting!'

Then Reinhart was coming forward, removing the shroud, winding the golden key. The performance began once more and Jeanne stared at the doll intently as it began to write. Surprised laughter from the audience, clapping, the usual repertoire. Reinhart had clearly been primed, as he took the note and began to make his way to Adélaïde. However, after a few paces he stopped, his brow furrowed. He read the note again, turned it over. 'I do not understand,' he said quietly. 'I do not understand.'

A murmuring gathered, rippling from the front of the room, and Louis stood and grabbed the note from Reinhart's hand. She saw his eyes flash like black diamonds. 'What is the meaning of this? Explain yourself!' His voice was low and dangerous.

'I do not know. This is not what I prepared her to write.'

Jeanne moved quickly to Louis and touched his arm, reading the piece of paper in his hand.

I was killed for what I know
But the truth speaks yet
The wrongs that I have witnessed
Will out despite my death.

'What is this?' Jeanne said to Reinhart sharply. 'What is the meaning of it? Is this some kind of macabre game?'

Reinhart's dark eyes darted about, panicked. He shook his head. 'I don't understand. I meant only—'

'Meant only what?' She looked about for Berryer and saw that he was already advancing towards them. She held out the paper to him, and saw his face harden as he read it.

'You will explain this,' he said to Reinhart. 'You will halt this little performance and you will account for what you have done.'

'It is not possible.' Reinhart grabbed back the paper, then held it up to his eyes, then began to shake his head. 'No, no. It was not my doing. What strange trick is this?' He looked about him, looked at the doll herself, as though wondering if she herself had somehow betrayed him. He lifted his hands in a plea. 'I did not ask her to say these things. You must believe me. This is not the message I asked her to write.'

The ripple of murmurs had now become a hum, the news of what the doll had written being passed from courtier to courtier, servant to servant, people putting down their wine glasses and muttering behind their fans, cheek to powdered cheek. And behind it was music, a song that Jeanne vaguely recognised.

'This is ... This is not possible,' Reinhart was stammering again. 'Someone else must have interfered. Someone else must have tampered with her.'

As he spoke, the gears of the doll started up again and he turned to stare, horrified, at his creation as she began – seemingly on her own – to write.

'What does she say?' a man shouted.

Others joined the cry. 'What does she tell us?'

Reinhart himself remained motionless and it was Richelieu who strode forward, took up the paper and read it aloud to the crowd:

'A child thief stalks the streets
Yet the police fail to detect them
Is it because the person we seek
Is under their protection?'

Berryer, white-faced, was talking to a guard in a doublet with slashed black sleeves. Two more guards were approaching them.

Then came a scream, a scream so piercing it might have shattered the mirrors about them. Then cries, shouts, the crash of glass and porcelain. People began to back away from the front of the room, towards the door. She saw then what they had seen. The doll was crying, as she had cried that very first time she was shown to the King. But it was not tears that fell from her eyes now. It was blood.

From that moment on, it was as though the time slowed, and Jeanne saw with terrible precision the falling crystal, the skid of pearl-embroidered shoes as people ran, slid, fell on the polished parquet flooring, their mouths open in shouts or cries. Berryer's arm was outstretched towards Reinhart and then came the surge of black satin, the tramp of men's boots as they ran to where the Doctor stood, his hands at his mouth like a frightened child. She could not hear what Berryer was saying, nor the men, nor could she hear whatever Reinhart was whimpering in protest. She looked away from them, away from the scurrying, screaming crowd, and at her. At the doll. The blood had streaked down her white cheeks and pooled in the lace at her throat. She was no longer beautiful, but horrifying. Her hand still held the pen that had written the message, but the desk before her was empty.

When Jeanne looked back at Reinhart she saw that he too was staring at the doll, a mournful look upon his face. There was something else there too, but she did not have time to read it before he was marched from the room between two red-jacketed guards, no doubt to the palace dungeons.

'Marquise.' Berryer ushered her, Louis, the Dauphin and the princesses from the room, past smashed glass and china, spilt wine and ruined food. An orange tree in a pot had been tipped over, its soil strewn over a gold-threaded carpet. A small dog was licking cream from a broken bowl on the floor.

'The doll,' she murmured to Berryer.

'It's to be shut away until we've worked out what to do with the wretched thing. Don't worry, Marquise. Leave it all to me.'

Louis, stony-faced, hurried to remove his daughters, Adélaïde being now quite hysterical, and Jeanne was left to make her way back to her apartments alone. As she rounded the corner at the Ambassadors' Staircase, she came upon Richelieu talking with a group of courtiers. He turned as she passed and held her gaze. 'New tricks indeed, Marquise. It was almost as if you knew what would happen. Almost as if you directed it.' The others turned to look at her too, their white painted faces as hard and pale as marble.

She continued walking past but felt their eyes on her back. Did Richelieu really think this was her doing? It didn't matter. This was not about what he believed, but about what he could persuade others to believe. She lifted her skirts and quickened her pace, hurrying to the safety of her rooms. She now recognised the song that she had heard: it was a children's rhyme, a lullaby, that Alexandrine used to sing. And it had not been sung by a member of the crowd. It had come from the doll herself.

24

Madeleine

Lefèvre insisted on taking Madeleine back to the Louvre in his carriage at once. 'I would advise, my girl, that you don't stay a minute longer. The King is extremely angry. Who knows what he intends to do.'

Madeleine hadn't known what to say to Lefèvre at first, not in fact been able to speak at all. She sat in the red leather carriage entirely silent, her mind turning over and over, the sound of the doll's singing still in her ears. Throughout these past days, her thoughts had been spinning themselves into a theory as to what had happened: Reinhart at the centre, creating the doll as a ghastly souvenir to the girl he'd destroyed; stealing children from the streets and attempting to rework them in metal and silver and glass.

But in the Galerie des Glaces she'd seen a man distraught, a man confused. Was it an act? He'd seemed genuine, but his whole career was about making the false seem real. Was Reinhart making the doll tell riddles that threw the blame on others – a clever trick like all his other inventions? If not, how had the doll done what it had done, written what it had written? And was what it wrote the truth? Anxiety worked at Madeleine's insides until she feared she might spew bile on the beautiful seats. For this wasn't only about Reinhart, nor about what was happening to the children of Paris, but about what would happen to her and what would become of Émile. She'd little doubt Camille would take the opportunity to blame her for not having seen what kind of man Reinhart was, not having somehow predicted what had been unleashed. Should she take Émile and run now, leave the city, before Camille could punish her for a failed spy? But where on earth would she go, and how would she support them? She still had no money, few possessions. Would she be able to secure a position outside Paris, with no reference and a ruined face? And more than that, she couldn't leave now. Someone had killed Véronique, seemingly the same person who was taking the children. She needed to know who it was.

For some time, Lefèvre left Madeleine to her thoughts. He himself looked agitated, his wig even more than usually askew. Only as they reached the outskirts of Paris did he speak. 'There will be a perfectly sensible explanation for what happened, and it will be confirmed that Reinhart had nothing to do with it. There is always some plot or intrigue afoot at court, you see, girl; there is always some argument. This will be nothing to do with your master, and it will all be resolved, I'm certain of it.'

She stared at him with his shiny face and beetle-black eyes. Did he believe his own bluster? It was difficult to tell. She'd met many men who were so sure of themselves and their paths in the world (that had so far been smooth and straight) that they couldn't picture things turning out badly, couldn't imagine how misery tasted. After a time, however, Lefèvre's face sank into a frown and he stared out of the window, and she realised he was trembling. She thought again of his horror when he'd first seen the doll, of his kindness to Véronique.

'Monsieur,' she said after a minute, 'does anyone else know how to work the doll?'

Lefèvre turned to look at her. 'Not to my knowledge, though I haven't spoken much to Reinhart of late.' A pause. 'It was his invention, really, though he worked with some of the plans that had been created by Vaucanson, and of course I assisted him with my own ideas. After Véronique's death, he insisted on keeping the details of the android to himself, not even sharing them with me. I confess I was rather peeved by that at the time, given this is very much my area. I thought he simply wanted all the glory for himself. Now of course I wonder ...' He trailed off. 'That is not to say others could not have watched him and worked out the mechanism for themselves. Admittedly, they would have had to have watched very closely. They would have had to be very clever.' He rubbed his face, thoughtful. 'But do not concern yourself with that. It's more important that you consider what you will do if this matter cannot be quickly resolved. Reinhart's household may be dispersed, you see, the apartments at the Louvre taken back. You have somewhere else you can go?'

Madeleine felt her stomach plunge. 'Not exactly, Monsieur.'

'Your family are not in Paris?'

'My mother is, but ... she would be unable to support me.' In all manner of ways.

'I see.'

'And so far as I know, neither Joseph nor Edme have anywhere to go.'

'Well then, we will have to do our best to make sure this is all quickly concluded, won't we? So that all can return to how it was.' He tried to say it as though it were the easiest thing in the world, just a wrinkle to be smoothed over, but she could see the anxiety in his expression, hear the note of strain beneath the bravado.

Madeleine tried to smile back at him, but felt herself to be on a tiny island, and all around her was the frozen sea.

By the time they arrived back in Paris it was late evening and the air was thick with the smell of burning lamp oil. The carriage wheels bumped over broken timbers, and as they neared the Louvre, Madeleine saw that the whole of the Rue Dauphine was full of broken glass, glinting in the lamplight.

'Such wanton destruction,' Lefèvre murmured. 'They are like a herd of stampeding cattle. How do they think such violence will help?'

'People are angry, Monsieur. The authorities do not act, and they have no other power.' And if the doll spoke the truth – if the child thief was being protected – then they were right to kick out in fury. 'In fact, Monsieur, I will dismount here. I need to go to my mother's house to check that they're unharmed.'

'No, no. We will drive you. The address?'

'You are very kind, sir, but it is not a good address. The Rue Thévenot.'

Lefèvre opened the carriage window and shouted up to the driver, directing him to change course. Within minutes they were approaching her old house and Madeleine insisted that they let her out when they were still some distance from the 'Academie'. As Lefèvre's valet helped her down onto the road she heard a shout and there before her was the little face she'd missed so very much.

'Madou!' he shouted, but he was staring not at her but at the gleaming coach with its twin chestnut horses.

She took hold of Émile's hand and looked back towards the window, nodding a thank you to Lefèvre. He touched his hat and the coach set off, Émile staring after it.

Madeleine bent down to him. 'You're all right? Lord, it's

good to see you. I'm sorry I couldn't come before. They made me stay.'

'At the palace?'

'Yes.'

'A real palace?'

'A real palace. I meant to bring you something back, but—' She wanted suddenly to cry.

Émile must have known as he put his head on her shoulder and let her hold onto him tight. 'There's been fighting round here,' he said to her, pulling back.

'Oh yes?'

'Yes, lots. Shouting and banging and people shooting!' He gestured excitedly. 'All sorts!'

'Sounds frightening.'

'Not for me. I was brave.'

'Oh, I'll bet you were. A brave soldier. Comfort any of the girls?'

'The new girl, she's nice to me, I looked after her.'

'Well, she's lucky, isn't she?'

He nodded. 'I'd rather have been looking after you.'

'Yes, so would I, Émile. So would I.' She considered running, then. Just taking him and what little she could scrape from the house and getting out of Paris before it took light. But where would they go? How would they survive with no friends, no property, no savings? She had a vision of them walking country lanes alone, ragged, hungry, like the many paupers who came to Paris hoping the streets were paved with gold, only to find the streets weren't even paved – they were just a river of stinking mud.

At the apartments, Edme took one look at Madeleine, led her to a seat by the kitchen table, poured a large glass of ruby red wine and set it down before her. 'Drink. Talk.'

Joseph took a chair close beside Madeleine. She could feel the heat emanating from his body. She took a gulp of wine and did her best to explain.

'It can't be true, though, can it?' Edme asked, her hands to her face, once Madeleine had described it all. 'That someone killed Véronique to silence her?'

'Well, it would explain why no one seems to have seen the carriage accident.'

'Yes. That I can believe,' Joseph said. 'But how to explain the rest of it? How to explain how the doll wrote this message without the master directing it?'

'It's witchcraft, is what it is,' Edme insisted. 'There's always been witchcraft in that court. They thought they'd burnt it out, but they haven't.'

She was talking, Madeleine knew, about the affair of the poisons. Maman had spoken of it often enough: babies' bones ground up to make spells and love potions, hundreds of people – both in the court and the city – arrested and tortured till they'd say just about anything; women burnt at the stake. 'That was a long time ago, though,' she said quietly. 'Under the old king.'

Edme shook her head. 'Not so long ago. Those practices were never rubbed out. Some say this king's mistress keeps herself young the same way the old king's mistress, Montespan, did. Did you know that?'

Madeleine thought of Pompadour's beautiful painted face, the too-bright eyes and flushed cheeks. She had, now Madeleine thought about it, seemed to grow younger, but surely that was impossible. 'People will say anything to do a woman down,' she said uncertainly. 'Particularly one with brains and power.'

Joseph was sitting staring at his hands. 'Where did they take Doctor Reinhart?'

'To the Versailles dungeons. At least that's where Lefèvre thought he'd been taken.'

'What will happen to him now?' Edme asked.

'I don't know,' Madeleine said, but she could guess. She'd heard plenty of tales about what happened in the prisons of Paris. She knew how they made people talk.

Joseph must have known too, as he stood up, suddenly decisive. 'We must find something we can use to defend him. We must show them Master Reinhart couldn't have made the doll do what it did. You will help me search the room, Madeleine, help me clear the master's name? You're the only one of us can read.'

Madeleine stared at him. Could he have no doubts about Reinhart, no fear of the man at all? 'Yes, I'll help you. Of course I will. There's just the small problem of getting in.'

They had to bring in the locksmith to get the workroom door open. Edme muttered and fretted about it, but Joseph spoke to her firmly. 'He can live with broken locks. What's important is that he does, in fact, live.'

Madeleine had half expected to see something horrific when they entered the room: discarded limbs and part-made machines, cogs and gears and eyes, evidence of Reinhart's other attempts at creating clockwork children. In fact, what they saw was merely a mess of tools and papers, evidence only of a scattered mind and a machine finalised in haste.

Joseph and Madeleine worked together, searching through Reinhart's papers. Joseph had only a little knowledge of letters, but he could interpret the many drawings and diagrams that they found in Reinhart's bureau. They sorted through the plans, lists and sketches, Madeleine reading out explanations of how the doll's lungs were filled with air, about how the doll could be programmed to write, and how a vial of water was used to produce the tears.

Then, in the top drawer of the bureau, she found a letter

Reinhart had drafted to the *Academie des sciences* explaining how he'd invented the world's first breathing, writing, crying, automaton. He wrote of how he'd hoped, in accordance with the King's wishes, to be able to find a way to pump blood through the doll's veins, so showing the circulation of the blood, but how he'd not as yet been successful. He wrote of how he had, 'despite numerous experiments been unable to produce a mechanism by which to produce authentic speech'.

Reading that line, Madeleine felt her own throat close. She'd heard the doll's voice herself and it'd been as true as a real girl's. She folded the letter carefully back into its envelope. 'We can ask Lefèvre to take this to Versailles,' she said. 'He will help us. It says that Doctor Reinhart didn't make the doll sing. If that's right, then it must have been something else.'

Joseph's eyes met hers. 'It sang?'

'Yes. A children's song, "Clair de la Lune".'

Prête-moi ta plume
Pour écrire un mot
Ma chandelle est morte
Je n'ai plus de feu.

''Least that's what I thought I heard, but it seems impossible.'

Joseph frowned. 'Much of what Reinhart did seemed impossible, but surely he'd have recorded it here had he achieved it. He'd have wanted everyone to know. Someone else is playing a trick and making the master pay for it.'

'Yes,' Madeleine said, 'it must be that.' But part of her was still wondering if it weren't Reinhart himself who was playing the trick, this letter simply part of the game.

As they left the room, Madeleine's gaze fell on the tray of coloured glass eyes that she'd first seen weeks ago. Now, however, many of the compartments were empty, the remaining eyes staring up to the sky.

25

Jeanne

After the incident with the doll in the Galerie des Glaces, Jeanne had taken to her bed, her head heavy, her throat excruciating, her skin damp with sweat. For the past two days she had kept to her rooms, her main company the monkey, Miette. Even there, however, the rumours swirled up to her like one of Paris's choking fogs: through du Hausset when she tended to her, through de Quesnay when he came to her with his medicine bag, through the Maréchale de Mirepoix when she came to visit.

'Some at court are saying, Jeanne, that the doll is operated by the devil. Can you believe that?' The Maréchale wore a dress of sunflower-yellow silk that made Pompadour's eyes water. 'They say she's continuing to write.'

'Anne, you know that isn't possible. Do close the curtains. My headache is monstrous.' She could feel its claws closing round her skull as she imagined the lies Richelieu and his entourage were spinning, the glee with which they'd paint her as a witch. *She claims to be lying on her sick bed, but all the while she makes the doll move.*

'Very well, but first you must see my new teeth.'

'Agate?'

'No. Real.' Anne came very close to Jeanne's face and bared her gums so that Jeanne could see the fresh white teeth that had replaced the little grey stumps. 'See?'

'*Mon dieu*. Whose are they?'

She shrugged. 'Some pauper who needed the money. I didn't see her – they extracted her teeth in the next room.'

A young pauper, Jeanne thought absently. The teeth wouldn't otherwise be so white. 'Very nice,' she said weakly. 'And now the curtains?'

Anne drew the drapes so that the room was enveloped in gloom.

'Tell me what you've heard about the news in Paris. Word has spread there, Hausset tells me, of the doll's messages about the children.'

'Oh yes, it's all too appalling,' the Maréchale said, her dark eyes shining.

'*Le bas peuple* have been rioting and breaking things in all different parts of the city. Apparently they're demanding that the doll be brought out before them so that she can speak the truth. Some are proclaiming that she's a modern-day Pandora, created to unleash vengeance on the city for the taking of the children. What do you think of that?'

Jeanne felt a sudden chill beneath the fever. She thought of the doll's silver garments, of the gold crown perched on its head. The monster's claws tightened and she put her hand to her head. Had the doll had been made to look like the Pandora of the myth: a messenger not of hope and joy but of an angry God's revenge?

'Well?'

Jeanne swallowed, seeing again the blood pooling at the doll's throat. 'I think this is the work of the clockmaker; that

he programmed her to write those messages and then, unsur-
prisingly, denied it.'

'But then how do you explain how she has *continued* to
move since the man was locked up in the dungeons? People
are saying she still writes, and that she gives other clues as well.'

'People say all sorts of things, my dear. They say I am dying,
that I am poxed. That I am a witch.'

Anne looked away from her, then said quietly. 'I don't see
why Louis doesn't just have the wretched thing destroyed.'

'Oh, I have proposed it, but to no avail. He wants first to
understand how it works. He thinks it is some kind of great
scientific achievement: true reanimation, something of which
the world will take note. But of course it's just a mechanical
doll, no different from that hideous doll the clockmaker's
daughter made for Alexandrine.'

Except that of course it *was* different. Whereas the reading
doll had been strange and unnerving, the Messenger, as he'd
called it, was something else: intended to cause harm.

'Well, I hope you're right about that. I must say the whole
thing has worried me terribly. And don't pretend it hasn't af-
fected you, because this is the worst I've seen you for months.'
She took out her painted fan. 'Does he still come and bother
you when you're sick?'

Jeanne breathed out but did not answer. She heard Miette
give a little cry.

Mirepoix fluttered the fan before her face. 'Ugh. If the
clockmaker was going to make a mechanical girl, he could at
least have made her useful, and given you a few nights off.'

Jeanne tried to laugh, but it was only a rasp deep in her
congested chest. For the first time in years, the King hadn't
touched her, had not called on her at all.

★

By the following day, her headache had abated, but her blood remained sluggish and the embroidered walls of her apartments seemed to press down on her. Instead of sitting at her toilette, Jeanne asked Berryer to accompany her as she walked about the fruit houses, where cherries and apricots scented the air, almost masking the stink of the palace.

'Well, Nicolas? Tell me what has been happening. I hear people are saying the doll continues to move, that she has given further clues.'

'There are claims of that nature, yes, but there is of course no truth in them.'

'Nevertheless, tell me exactly what is said to have happened.'

Berryer rubbed his jaw. He looked, she thought, more exhausted than she'd ever seen him, his skin candle-pale, but then she was hardly in looks herself. 'A servant, upon going into the room where the doll was kept, insisted that it had moved from where it had been set down, over towards the window. This must, of course, have been a lie or a misunderstanding, but after that, the rumours have continued to accumulate. Writing was found on the wall of the room in which the doll is locked, which must of course have been a forgery, but—'

'What did this writing say?'

'That she will not stop until the child thief comes forward, or something along those lines. I forget the exact wording. It was clearly nonsense. In any event, the doll's pen and ink have been removed, so there will be no more of this message writing.'

Jeanne felt strangely cold. 'What else?'

'Honestly, Marquise, it is all—'

'What *else*, Nicolas?'

He passed a hand across his brow. His voice took on a resigned tone. 'Two serving women passing the room claim that they have heard the doll singing. Another boy insisted the room smelt of urine as though the doll had passed water, but as we know, that is how the whole of Versailles smells. In any event, the doll was carefully inspected again this morning by men from the *Academie des sciences* who concluded that, as Doctor Reinhart had said, it can only write what it is programmed to write. There is no other way of operating it. It cannot have designed its own messages. Moreover, it cannot possibly have moved of its own accord. It must be wound in order to operate.'

Except that the doll had written two messages before her very eyes, the second apparently on its own. She picked a peach and held it to her cheek; it was softer than her own skin. She wanted desperately to see Alexandrine, to hold her, smell her hair, that scent of milk and soap. She wanted to bring her back here, but then was it safer here than in the city? 'The Maréchale du Mirepoix tells me that the people of Paris are clamouring to see the doll, claiming she's some oracle or harbinger of revenge for the taking of their children.'

'Yes, there has been some of that. Madness, of course, but the people of the Third Estate have always been excitable.'

Particularly so when their children were going missing. 'There is full-scale rioting?'

His expression was pained. 'There are some difficulties, but they are being suppressed.'

'And are you any closer to making arrests? Of the child thieves?'

'Not of thieves, but of subversives.'

'Oh?'

'Yes, we have reason to be believe that the allegations of vanishing children are in fact the work of a gang of immigrants wishing to stir up trouble against the Crown.'

'Really? Who are these people?'

'Malcontents from outside Paris. My officers have identified a handful of possible culprits: men who have tried to whip up mayhem before. One was heard offering money to labourers to make claims that my officers were spiriting children off to Versailles for to provide blood for,' – and here he dropped his voice – 'the King.'

'Good God. Have you told him?'

'Not yet. We are gathering further information.'

'Why would these people say such a thing?'

He shook his head. 'It is difficult to know. We fear it may be some plot to undermine the state – to make the people rise up against their king. They will not succeed, of course. The guilty shall be identified and dealt with harshly. But we must consider the possibility that Doctor Reinhart – himself of course Swiss – is one of them, and that the messages written by his doll were part of this very same plan.'

'Must we?' Jeanne found it difficult to reconcile the vision of the thin, peculiar-looking clockmaker with that of a radical who sought to stoke anger among the populace. 'But she found nothing did she, your *mouche,* to suggest Reinhart held any controversial ideas? Nothing to suggest he was a troublemaker of any kind?'

'No. Evidently she has not performed her role as she should have done. I should have vetoed her appointment once I knew … She will be spoken to.'

'Knew what? That she was a prostitute? Surely so are half the spies on your books.' Jeanne shook her head. 'No, I'm not convinced of this theory. The girl seemed shrewd and the clockmaker hardly seems an anarchist.' She was tired now. She took a seat on one of the wooden banks. 'The doll needs to be destroyed. Before further damage can be caused and further rumours spread.'

'I agree, Marquise, entirely, but that of course is not up to me, and you have more influence in that quarter than I do.'

Not anymore. 'I will try again, but so far my attempts to convince Louis have failed.'

'Perhaps you'll have more success in convincing him to have the clockmaker put on trial.'

'Louis refuses? Why?' she asked sharply.

'He says it will achieve nothing, given Reinhart continues his denials. Perhaps he feels some ill-placed pity for him.'

Unlikely. Louis rarely considered the feelings of others, for in all his life he'd never had to. Since birth he'd been surrounded only by obsequious courtiers, not people who told him the truth.

'What has the clockmaker said so far?'

'Very little, despite being put to the Question.'

Jeanne winced inwardly, imagining the straining and splintering of bone. 'He continues to deny his involvement?'

'Yes, he still maintains he was not responsible for the messages, which means he's either insane, or of stronger stuff than we'd imagined.'

'Or he is telling the truth.'

Berryer turned his lips down as if in doubt at this suggestion.

'And are there any other rumours or theories? About the doll? Or about the children?'

'None that are worth talking of.'

She smiled thinly. 'Nicolas, you know that it's best to tell me everything. That I like to be apprised.'

He shifted in his position. 'As always at Versailles there are claims of black magic, ones which have no basis in fact.'

'Claims against me, you mean. Do be plain.'

'I regret so, Marquise.'

'The specifics?'

He grimaced. 'There was a further *poissonade*, intercepted before it reached your rooms. I had not wanted to disturb you with it given your illness, but it claims that it is you who operates the doll in order to falsely accuse the King and detract from your own … activities.'

'Those activities being?'

'Using the blood of the stolen children to restore your own declining health.' He paused. 'Ridiculous, of course.'

Jeanne closed her eyes. A blackness rose before her closed lids and she felt that she might faint. 'Yes, indeed. Ridiculous. Does Louis know of this latest *poissonade*?'

'Unfortunately it was brought to his attention.'

Cold fingers of fear crept up her spine. Was that why he hadn't visited her? Had he begun to believe the rumours were true? 'You will identify the authors of this delightful verse.'

He nodded. 'It is already my priority.'

'Evidence, Nicolas. I need evidence, or the King will refuse to act, as he has so often before. And you will come up with an explanation as to what has really happened to the children.'

'Marquise, I assure you we are doing our best.'

'Your assurances are no longer enough, Nicolas. This whole sorry saga needs to be shut down, just like that wretched doll. It's not good for my health, nor for the King's reputation. Nor for your career.'

★

In the centre of the ceiling, a white horse drew a chariot across the sky, the sun god himself within it, shining down on the courtiers of Versailles. The Apollo Room was the grandest of all the drawing-rooms at Versailles, every inch coated in silver or gilded wood or gold-embroidered velvet, or the most exquisite paintings. It was a shrine to Louis XIV, the Sun King. Her Louis therefore hated it, for it threw into the light his own inadequacies as ruler.

From among the throng the surgeon, Lefèvre, emerged,

his face unpowdered, his wig askew, his lace cuffs distinctly grubby. Jeanne wished the man would take more pains with his dress, particularly given the pain her own dress caused her. He approached the throne and gave a low, inept bow, from which she feared he would not recover himself.

'Claude,' the King nodded. 'Today is not our day for lessons, is it?'

'No, sire. It is not.' He was perspiring, she noticed, a sheen of moisture across his brow and upper lip. 'I come in fact to entreat you to allow me see my friend,' he lowered his voice, 'Doctor Reinhart.'

'Monsieur, I had thought better of you. There is nothing you can say—'

'Please, sire.' The man knelt on one knee, so that he looked unbearably ridiculous. 'Hear me for merely a minute. You trust my judgment, I think, in matters anatomical, physiological. I hope you will trust me on matters of the mind as well.'

'You have taught me well, Claude. But I do not wish to hear your views on this matter.'

'Perhaps not, sire, but I have had word that Reinhart is being badly treated, in a very poor state, and I will never forgive myself if I make no application on his behalf. It is vital that I see him.'

Louis breathed out. 'You have one minute.'

'Thank you, Majesté. I wish only to say this: that the man has always had his peculiarities, it is all tied up with his particular genius, but I do not believe he was responsible for the messages written by the doll. It must have been someone else – someone else who worked out how to operate the mechanism and who wishes ill upon us all. It is *that* person you should be looking for.' He looked panicked, she realised. Did he think he himself would be blamed? 'Majesté, consider this: there is no reason for him to have done it. None at all. It merely put him in jeopardy. As for the claims that the doll has somehow moved or spoken since it was shown, I can only say that Reinhart does not have the power to animate it in such a way. It is beyond his capabilities, indeed beyond those of any man. I show you here his letter to the *Academie des sciences*, as found by his servants, in which he confirms that in fact he had failed in the task you had set him: he had been unable to find a way to make the blood circulate or make the automaton speak. Those were not things he did.' He held out the letter. 'I believe him to be innocent, sire – the victim of some plot or trick. I beg you to have the doll destroyed and to allow me to see Reinhart. The police will allow me no admittance.'

The King did not look at Lefèvre, nor the letter he held out, but at the fireplace of blue-grey marble. After a moment he said flatly, 'I do indeed trust your judgment in matters anatomical, Claude. You are a good teacher. But you are limited in your abilities. As you say, Reinhart is a man of strange genius. Whatever he claims in this letter, I think it is quite possible that he has animated the doll such that it can continue its machinations despite his being underground, shackled to a wall; despite the scientists at the *Academie* claiming the doll cannot operate on its own. Until he has explained to me how he reanimated his daughter, he will stay there and the doll will remain locked away. You may leave me now.'

Jeanne saw Lefèvre's face whiten. The sparkle which she had previously seen in his eyes had now quite gone. He folded the letter back into his waistcoat, gave a brief bow and walked stiffly from the room.

For a few minutes after Lefèvre had gone, Louis and Jeanne simply watched the courtiers in their glistening dresses, all trying to catch his eye. Jeanne glanced at Louis, noting that his jaw was still set. There was little point in speaking to him now.

If he could casually imprison and torture a man with whom he'd worked for weeks, she had little doubt what he would do with her if she yet again displeased him.

After a while, when she judged his anger had abated, she ventured smoothly: 'Surely it is time to get rid of the doll, *mon amour*? I know you wanted to find out its secrets, but it's been inspected twice now. This is all clearly some trick. Why not have it destroyed? It has caused so much distress and provoked so much unnecessary rumour.'

Louis shook his head. 'I have told you before,' he snapped, 'and I will not tell you again. I wish to know how it operates. I wish to know how a man has made an apparently inanimate object write and cry and bleed and sing. I want to know how he brought her back to life and I want him to teach me the trick of it.' He looked at her then. 'Reinhart might be a devil, but he is a devil who's achieved more than Lefèvre and the whole of the *Academie des sciences* put together. If he can do that, what else might he be capable of? How else might he advance science and the cause of France itself? He will talk in time, I do not doubt it. And he will talk to me.'

'But—'

'Would you prefer, my dear, that I believe the other theory that is spoken of ever more loudly at court? That it is my own mistress that makes it move? The mistress who despite her advancing years grows ever brighter-eyed?' He turned at last to look at her. 'You must be very careful, *ma petite*. You walk on broken glass.'

*

There was a guard sitting outside the room, half asleep. He straightened as Jeanne approached, and Jeanne put her finger to her lips.

'Madame,' the young man whispered, 'I was told to ensure no one went in. The door is locked.'

'My dear, I am not "anyone", as I'm sure you know. Which is why I have a key.' Berryer's spies were useful little fellows.

'But Madame—'

'It would be unwise, I think, for you to continue speaking, or indeed to be here at all. If you value your position you will take this moment to pay a visit to the necessary house. It may take you some time to return.'

Jeanne felt a rush of cold air as she pushed open the heavy door. Inside, the room was dark as jet, lit only by a candelabra, the flames of its candles casting shadows about the walls. And then she saw it: there, in the far corner, stood a tall, human shape. Jeanne drew in her breath and approached slowly, collecting one of the candles, straining to hear past the sound of her own blood. Forcing herself to act before her courage failed her, she pulled the velvet shroud from the creature's head. The face was a pale moon in the semi-darkness, its glass eyes a shining black. There, that wasn't so bad, was it? The doll looked much as it had when she last saw it – its expression blank, its eyes dead. The blood had been cleaned from its face, however, and its pen and ink were gone. Its crown too had been removed.

'I see you've been dethroned, *ma poupée*.' Jeanne could almost laugh at herself now, for being so afraid of something made of screws and discs and springs. The doll was certainly bizarre, but, standing motionless, it was very obviously a machine. Jeanne walked still closer to it, annoyed with herself for having waited until now to seek out the doll. It was, after all, merely a large toy, made for an oversized child.

But as she came within two foot of the doll there came a rustling sound, a whispering that seemed at first to emanate from the walls, but which she realised with an icy horror must be coming from the doll herself. She began to back away, her breath coming too quick, too shallow, her heart roaring. But above it she could hear what the girl was whispering:

'*Who was it, Madame, that wanted me gone?*
Who was it wanted to hurt me?'

26

Madeleine

In her hand, Camille's note. Outside her window, the crowd, rustling like a swarm of angry insects. A creature was growing in Paris, made up of the pressed-down furies and injustices and deprivations of a people who would no longer be silenced. In the Quartier Saint Germain de Pres, hundreds of men and women had gathered to demand that the doll be brought before them, to tell them the whole truth. The Watch, the people said, were concealing the crimes, even taking the children themselves. No one in authority would tell them the truth, but the clockwork girl would speak: she would reveal to them the true role of the King, bathing in their children's blood. On the Porte Saint-Denis, a crowd of thousands had risen up and smashed the police offices, then rioted in the Rue de la Calandre, leaving the city in splinters and smoke.

Despite or because of all this – the chaos, the bloodshed, the fear – Madeleine had known the police would summon her, and she knew what they were going to say: that she'd failed to work out what Reinhart was about and what he and the doll would unleash; that they'd give her one last chance to atone for what she'd done. But beneath her usual fear was anger and something more besides. Perhaps it was the rage that burned through Paris, or perhaps it was the anger that had been building up within her chest for years, but she wasn't having it anymore. None of this was her fault. The police owed her the money they'd promised her long ago. God knows, she'd more than earned it. She was tired of being told she was worth less than nothing by men who did nothing themselves.

The note told her to come to an address to the north of Les Halles and to leave as soon as she could. Madeleine wrapped herself in a shawl and stepped into the molten wrath of the city. It was the worst she'd seen it so far. While many shops and houses were boarded up, others had already been smashed – broken bricks and pieces of wood lying among the shattered glass. Men and women had taken up pieces of piping or wrenched steel bars off gates to brandish at any member of the Watch who dared try to disperse them. By the time Madeleine reached the Rue Saint-Honoré, her skin was damp with fear. Outside l'Église Saint-Roch a clot of people had formed around a man and were grabbing at him, hitting him with stones, tearing at his hair. 'Child thief!' they were shouting, 'Police spy scum!' 'Tell us where the others are!'

She could only glimpse the man through the press of bodies, but what she saw was a wreckage of blood and bone, and she felt the cold breath of fear as she imagined what would happen if they discovered her for a police *mouche* too. Pulling her shawl up to cover her face, Madeleine tunnelled and shoved her way through the crowd of people, pushing past bystanders and weaving through groups in a state of increasing panic.

The Rue de la Tixéranderie, the address given on the note, was a narrow street where the houses leaned into each other like drunkards. Protruding from the windows were long rods on which grey washing hung, and oversized shop signs that creaked in the breeze: an immense boot fashioned from steel;

a quill and inkpot designed for a giant. The street was empty, but from not far away she could hear the sound of a fight. She checked the piece of paper once again and then knocked on the wooden door. A drab-looking maid answered and showed Madeleine up the stairway to a room full of the sour, fleshy reek of tallow candles, where Camille stood alone, a glass tumbler in his hand.

'Well, well. Here creeps the little fox.'

Madeleine looked about her, at the damp-stained walls and the scorch-marked lamp. She'd imagined him plumper in the pocket than this. Were they not paying his wages either?

'What do you think, then? Tell me your theory. How is he managing to do it?'

'I've no idea. No more than you have. I don't even know if it's Master Reinhart who's behind it.'

From outside came the sound of shouting, closer now, the shivering of glass, the splintering of a door or table.

Camille walked to the other side of the room, his seedy frockcoat seeming too large for his shrunken frame, and poured himself another measure of brandy. 'Oh, it's him all right. It must be. *Cher Dieu*, how can it have escaped your notice that the man was entirely insane?' He stopped in front of her and she saw that he wanted shaving, his face grey in the lamplight. 'Can you imagine how this reflects upon me? Paris runs mad because of what that man is doing, what he claims. There's talk of rising against the King! Your mother told me you were clever. Bright as a button. Yet you claimed to have found nothing against the man; you gave him your seal of approval!'

Madeleine felt anger rip through her. 'I did no such thing and you know it. I only said I'd found no proof he was dangerous, which was the truth. I did everything you asked of me. More. But I could never have known that the doll would write what it did or do what it's done. You don't even know that this is Reinhart's doing!'

'What other explanation is there?' His voice, she thought, was sharp with desperation.

'Well, couldn't it be like Reinhart said: that someone else meddled with how it works? Or—'

'Who? Who *is* this person? And it is also they, this mysterious other, who makes the doll sing, who makes it whisper?'

'Whisper?'

'The doll continues to speak. It continues to give ridiculous messages.'

'What messages?'

'That isn't your concern. What *is* your concern is that your damned master the clockmaker is somehow managing to operate that ghastly doll despite being moved to the hell-pit that is the Châtelet dungeons.'

Why had they moved him? To put him on trial? Or to get him further away from the doll? Madeleine swallowed. 'Where is the clockwork girl?'

'She should have been destroyed – torn to pieces – but she's locked in a room at Versailles, heavily guarded. And yet still she speaks. Explain that.'

Madeleine imagined the doll locked away, her eyes glistening in the torchlight. 'If the girl still moves even though Doctor Reinhart's in gaol, doesn't that mean that someone else is working with him? Someone who could gain access to the palace?'

'Have you found some evidence of this?'

'No, but it's the only real explanation, isn't it?'

'Have you searched the Louvre? Tried to find some clue as to who this accomplice is, as to what Reinhart was planning?'

'Yes! I've searched everywhere.'

'You must have found something.'

She shook her head. 'Only letters and plans that show he hadn't managed to get her to speak or bleed. I've found nothing to explain it, nor link it to anyone else.'

Camille slammed his glass down onto the table. 'Then you're not looking hard enough. Or you are protecting him. There must be something to show that he planned this, and that someone else is involved. There must be a way to stop the thing, to shut its stupid mouth.' Camille stopped pacing and looked at Madeleine, though he seemed to stare right through her. After a moment he said, 'Does Reinhart trust you?'

'I'm not sure.' She thought of how he'd looked at her before they went to Versailles; of how he'd told her he had a reason for doing what he did. 'Perhaps.'

'Yes.' He rubbed his face, slowly. When he spoke again his voice was more measured. 'This is what you're to do. Listen carefully, because everything now hangs on it. You will go to visit your master in his dungeon. You will tell him that you'll help him to escape, but only if he admits to you how he is operating the doll, tells you who is assisting him, and agrees to shut it down.'

'But I won't be able to help him escape, will I?'

'No.'

'Then it's a shabby trick, and I won't do it.' It was more than that, of course. The truth was that she was scared to go there, frightened to see the man who she still didn't know for a murderer, a conjurer or a dupe. Nor had she any wish to enter the cells that were notorious for their miasma of disease and their legion of sharp-toothed rats.

'It's a little too late to grow a conscience, Miss Chastel. You were all too happy to accept the job when I first proposed it. You will do this and you'll succeed. You will make Reinhart tell you, or he'll remain there until he dies.'

'No.' Madeleine felt rage bubbling within her, threatening

300

to burst from the surface. She'd done what she'd been asked to all her life, she'd lied and deceived, even when it hurt her. But it would never be enough. They would always ask for more. 'I won't do anything more for you. Not until you pay me what I'm owed.'

'You won't?' He sat down at a table and picked up a piece of paper, then began to write.

'What are you doing?'

'I am completing an *ordre de cachet*.' He carried on writing, blew on the paper, then held it up, so that, in the top corner, Madeleine could see her own name in glistening black ink. 'It's one of the few perks of this wretched position, you see: I can lock up whoever I choose, and only the King can release them. And what will your little nephew do then, Madeleine, if you're kept year on year in the Bicêtre?'

Madeleine regarded him for a long moment: his sunken eyes, his unshaven face. 'You know what Suzette used to say about you? That you were a pathetic little man with a shrivelled little *quequette,* who dealt out pain to make himself feel bigger. That most of what you said was lies. I should have remembered that. I should have told you stick your wretched job up your lying *connard.*'

She thought he would hit her then, but he no longer seemed to have it in him. 'Well, you didn't. You were too fixed on the money – on your brilliant future – to think about that, and your little bitch of a sister is dead. So now here we are. What's it to be? A brief trip to the dungeons, or a long one to the Bicêtre? And if you come back to me empty-handed, I'll find that little nephew of yours and flog him half to death.'

Madeleine regarded him steadily. 'This is the last thing I'll do for you, for any of you. If I can't get Reinhart to tell me how he operates the doll it'll be because it isn't him operating it, alone or with another. That it's something else altogether.'

She saw it written clear in Camille's face, then. He was afraid. For all his claims and threats and *lettres de cachets*, he was just a little boy, afraid of being left in the dark.

They went at once to the Châtelet dungeons, Camille leading Madeleine down the Rue de la Coutellerie where a group of workmen were clashing with officers of the Watch, then onto a series of narrow, stinking roads, down towards the river and into the prison compound. In a cold, stone reception room a robed guard waited, his face white and bloated as a fish's. As Camille muttered to him, the guard passed his amphibian eyes over Madeleine, then nodded. He picked up a lamp and walked towards a wooden door, looking briefly back over his shoulder to indicate Madeleine should follow. She refused to look at Camille, only followed the man and his flickering lamp, the door swinging shut behind him. The man led her down a dark passageway that smelt of damp and misery, then down a stone staircase and along more narrow passages, the smell becoming more pungent, the light ever weaker. Sounds seeped through the walls: distant wails, shouts, the occasional scream of pain. Madeleine's cloak was thin and she was so cold by now that she'd near lost the feeling in her fingers.

At last, in the depths of the prison, the guard stopped at a heavy wooden door, and opened a little grille to peer inside. Then he unbolted and unlocked the door, which, when it was opened, whined like some strange beast. Inside, the smell was foul as hell: of vomit and piss and mouldering straw. The guard lit a candle in the wall sconce. 'I'll return in a quarter hour.'

In the semi-darkness, Madeleine could make out what she thought might be the curve of a prone figure in the corner of the room. Walking closer, she could hear breathing, rasping and quick. She put out her hand, which touched rough covers, maybe sacking. 'Doctor Reinhart.'

A moan. 'Véronique.'

She drew back. 'No, Monsieur. It's Madeleine, your maid. Come to see how you are.'

A croaking sound and then the figure shifted. 'Well, it seems I am still half alive.'

She moved closer again, and could smell his breath, his empty stomach and his sickness. 'Are you very ill, Monsieur?'

'I believe so, yes.'

She touched his head, and it was as though the flesh was burning. Her eyes having adjusted a little to the gloom, she saw a pitcher of water and helped him put it to his lips, heard him take a deep gulp.

'They've hurt you bad.'

'It was to be expected. They think me a heretic. Or a magician.'

She helped him lie back down. 'But who did it, Doctor Reinhart? Who made the doll write that message? Who made it sing like that?'

No answer, simply a rasping sigh. 'How did you get in, Madeleine?'

'I know one of the guards. I begged him. I'd heard that you were sick.'

Her guts twisted at her own lies, but if she couldn't get him to answer it could be her lying here, festering in the rotting straw. She tried again: 'They're saying that the doll is speaking, whispering. That it's still giving clues.'

A throaty rattling sound came, which she realised was a laugh. 'Yes, they say that. That is why they broke my arm.'

She closed her eyes, gritted her teeth. 'But it can't be you, can it, Doctor Reinhart? Because you're here? So who is it? Is someone else moving the doll? Do you know? Do you know who's taking the children?'

Silence, apart from the laboured breathing.

'Monsieur,' she whispered, 'if you can tell me, it may be that I can get you out, that I can convince the guard I spoke about. Is someone helping you, Doctor Reinhart? Is someone else working the doll?'

He gripped her hand then in his feverish palm, and it was all she could do not to draw it back. He pulled her close to him so that she breathed in the rank smell of him and she had to swallow down her bile. 'They thought they could silence Véronique, but still my brave girl speaks.' The throaty laugh again, morphing into a groan of pain.

'You are very ill, Monsieur.'

He had collapsed back onto the straw, his breathing heavier. 'Yes, I have gaol fever and perhaps poisoned blood. I told them everything I knew, but that, it seems, was not enough.'

Seeing his mangled, tortured body, Madeleine was seized with a violent guilt. She'd helped get him into this situation; she'd spied and lied from the very start. 'They can't do this to you. You're the King's clockmaker!'

'My life is no more valuable than anyone else's, girl. In the end we all return to dust.'

'What would heal you? What would make you better?' She could go to Lefèvre. He, surely, could help.

'Perhaps nothing at this stage, Madeleine. I am dying, I think, but Véronique lives on, the spirit in the clockwork doll. Stronger than any of us, than I could ever have hoped.'

He was mad. Quite mad. They'd fractured not just his body but his mind. And Madeleine felt a stronger rush of guilt as she recalled all the things she'd thought him capable of. He was not a murderer, not a demon. He was a broken and dying man. 'Doctor Reinhart, I will get medicines for you. Please, try not to despair.'

'You? You will help me!' He laughed.

A stab of fear. Did he know? Had someone told him? Had

he guessed what she was, how she'd really got in? Or perhaps he was just laughing at the idea of being saved by a servant girl.

A gust of cold air as the guard opened the door. 'Time to go.'

Madeleine's gaze didn't leave Reinhart's face. He was mumbling to himself now, smiling. She twisted around to look up at the guard. 'You can't keep him like this. You can't let him die here, or they'll never know the truth, will they? They'll never find out what happened.' She stood up. 'He needs a physician.'

'Out. Now,' the guard said, unmovable as an oaken door. 'You shouldn't be here at all.'

Madeleine bent down next to her old master. 'I'll do my best, Monsieur. I'll do what I can. I'm going to your old friend.'

*

When she made it back out onto the street, Madeleine doubled over and retched, though her stomach'd been nearly empty. Leaning against a wall, she gulped in the Paris air that reeked of shit, but which was still a hundred times healthier than the foulness back in prison. She had to leave Paris. There was nothing for it now. Madeleine didn't doubt Camille had it in him to half kill Émile and lock her up into the bargain once he found she'd failed to get Reinhart to talk. She wasn't sure how she'd keep them alive if she ran, but to stay here would be worse.

She hurried from the Châtelet compound onto the Rue Saint Germain, where the air smelt of gunshot and burning straw; in the distance, shouts of fury. She'd go first to the Louvre, to pack up her things and say goodbye to Joseph. It was acid burning at her insides, knowing she must lie to him. Knowing she must leave him. But what other choice did she have?

'Where will you go?' he demanded of her once she was back

in her room, cramming clothes into her case. 'How will you survive? Just tell me, Madeleine – tell me what's going on. It may be that I can help.'

'I can't, Joseph. And you can't. I'm sorry. I'm sorry, truly, for everything. I have to go to my nephew, then to Lefèvre. I have to beg him to help Master Reinhart.'

'You think he is dying?'

'I know it, Joseph – you should've seen him – and I can't think of anyone else who'll help.'

'Lefèvre has tried to get to him before. They will not let him in.'

'Then he'll need to try again. There's nothing else I can do.' She continued with her packing but was aware of Joseph's eyes still on her. She looked up and was suddenly seized with a ridiculous urge to tell him: of everything she was, why she was, what had happened to her. But of course she couldn't. She heard her mother's voice like a wasp her in her head: 'Keep it shut, girl. Whatever he might tell you, there's no man wants to hear it.'

She could explain little more to poor Émile, whom she tried to hurry to collect his books and blanket so they could leave her mother's house. Camille would be looking for her by now, she guessed, and this was the second place he would come.

'But where are we going, Madou? Why won't you tell me?'

'Émile, please keep your voice down, my pet. I'll tell you everything when we're on our way, but right now we need to keep our minds on the things we need to take.'

Carrying their cases, Madeleine dragged Émile along the Pont au Change and onto the Île de la Cité. As they turned the corner into the Rue de la Calandre, she saw a group of people gathered outside a tall house, battering at the door and throwing stones at the windows.

'There'll be no mercy for those that hide the bastards. Open this door up now!'

Fear sliced through Madeleine like a cold blade, fearing they were talking again about police spies, the men they believed were taking the children. She pulled Émile on more quickly.

All was quiet on Lefèvre's street, but the windows to many of the houses were shuttered against rioters and though she rang the bell and banged on the door, still no one came.

'Please!' Madeleine shouted. 'I come begging help, for a friend of Monsieur Lefèvre's!'

After a moment a valet opened the door a fraction and appraised her through the crack with pale and watery eyes.

'Please, Monsieur,' she said. 'Tell your master that Madeleine, Doctor Reinhart's maid, is here and very much needs his help.'

After a few minutes she and Émile were shown up to a parlour where a fire burned brightly, throwing its light on rich furnishings: lacquered wood and ornate tapestries, and where a very old woman sat sewing. Lefèvre stood up as Madeleine entered the room and she saw that his face had grown thinner, gaunter. He wore a quilted dressing gown over his suit.

'Madeleine. What brings you here? Ah, your little nephew. Is he ill?'

'May I speak with you alone, Monsieur? It won't take long.'

'Of course.' He glanced at the old woman, still sewing with tiny stitches, seeming not to notice their visitor. 'Maman, I'll be back shortly. Mademoiselle, come with me.'

He led her and Émile through a door at the back of a room and down a spiral staircase into a dark-wood panelled office. Madeleine could see specimens in jars like those Reinhart had in his workshop, and anatomical sketches on the walls. Émile stared at them in wonder as they passed.

Lefèvre lit sconces in the walls and led them to some leather chairs. 'Tell me, Madeleine.'

'Monsieur,' she said quietly, 'it's Doctor Reinhart. I saw him in prison.'

'How is he?'

She glanced at Émile, but he wasn't listening. He was staring instead at the jars of body parts, a skull that had been eaten away.

'He's terribly ill,' she whispered to Lefèvre. 'Gaol fever and infected blood, he says. Several bones are broken. They crushed his legs.'

'*Mon Dieu.*' He rubbed his forehead. 'I'd heard that he was ill, but this is execrable, inhumane. How did you get access to him?'

'A friend,' she lied. 'A guard. I begged him to let me see him, but they won't let me in again.'

'Well, you have done very well, Madeleine. Very well indeed. I tried to get to him myself but failed. I should have tried again. I should have tried another way. I can get no answer from my friends at the palace as to what on earth is going on.'

'Monsieur, I beg you to try again, before it's too late. Go to the Châtelet direct and tell them he's dying. They'd not listen to me.'

'I will do that. I will do that at once. Let me mix some medicine for the fever, some ointment for the breaks. What else, what else?' he said to himself, standing up. 'I'd ask Marthe to prepare some restorative food to take but all the servants have already left. Indeed, it's lucky you came when you did, for I leave Paris tonight.' He shot a look at Émile, who was coughing, taking in his meagre frame. 'Perhaps I will mix up something for your nephew too?'

'Monsieur, I'd be very grateful. His chest hasn't been good for a long while and no medicine I've tried has worked.'

As she waited, Madeleine stalked up and down the corridor, looking absently at the preparations in jars: a heart, perhaps, preserved in spirit, a ghostly floating hand.

'Why do they keep them in jars, Madou?' Émile was looking intently at bloated white slices of something labelled 'cerebellum'.

She rubbed her apron over her face. 'For teaching, my love. They preserve things so others can learn what bits of the body look like, what they do. That's what Véronique told me at any rate.' She thought back to Véronique explaining preparations to her the first day she'd come to the clockmaker's, showing her the moving spider. Madeleine had thought her manner faintly superior – yet another person trying to put her in her place – but Véronique hadn't been that at all. She was only lonely.

And then those thoughts vanished as she saw, on a low shelf, an ear preserved in a jar – the skin dark, a hole pierced in the lobe. For a moment time seemed to freeze as she pictured Victor: the small silver hoop he had worn in his right ear, the childish, mischievous face.

'Here we are, then.'

Her blood was ice. Lefèvre was standing just behind her and, as she turned, his eyes fixed on hers, fathomless and dark. 'I've mixed up some cough mixture for your nephew too. You must bring him back to me when there's more time.'

Madeleine could barely breathe. Had he seen where she was looking? Had he guessed at what she'd thought?

'Thank you, Monsieur. You're very kind.'

'Not at all, not at all. Now, you'll forgive me for chivvying you out, but I must go to the prison at once.' He led her back up the staircase, talking of his efforts to gain access to Reinhart, his concern for his welfare, the King's refusal to listen.

But all the time, Madeleine was thinking of the ear behind glass, and wondering if Lefèvre had seen what she'd seen, knew what she'd thought. She was acutely aware of how close Lefèvre was, how pale and dry his skin, how black his eyes.

She could see the dusting of wig powder on his shoulders, smell the scent of the liniment still on his hands. She had to get them to the door. She had to get out of this house.

They had entered a long corridor with a carpeted floor and dark portraits lining the walls. Émile was lagging behind them, staring up at a painting of a young boy holding a dog. She turned to him, stretched out her arm, opened her mouth to call, then felt the air move behind her, the shock of the blow, and everything was blackness.

27

A clock was chiming. From nearby, the whirring of some machine. Slowly, painfully, Madeleine clawed her way back to consciousness. Her throat was dry as dust, her mouth furred. When she opened her eyes it was to a wavering half-light, a lamp burning in another part of the room. How late was it? Midnight? Later still? The cogs of her mind moved slowly and as she struggled into a sitting position, her skull screamed out in pain. She couldn't, however, put her hands to her head, as she found they were bound together. So too were her feet, which were strapped, she realised, to some kind of worktop or table.

A wave of nausea passed over her, then the light brush of terror as she saw on the other side of the room another worktable, another figure, one much smaller than her: Émile.

'*I'll look after him,*' she'd told her sister, '*I'll keep him safe.*' And she'd brought him here to a madman's lair.

'Awake so soon?' It was Lefèvre, now wearing an apron over his suit. 'I always thought you were rather too quick, rather too knowing for a maid. Shouldn't have looked where you weren't supposed to, should you? Well, there it is. Here we are.'

'My nephew,' she managed to say. 'Please let him go. He's only a child. There's no reason to kill him.'

'Well, there's little reason for him to live, is there?' Lefèvre said casually. 'An orphan, or as near as. What will he do once you're gone? His life will be miserable and pointless, as for so many in this city. Through his death, however, he will serve a purpose. For the first time in his life he will be valuable. He will be part of the advancement of science.'

Science. Was she awake? 'What science?'

'The science, Madeleine, of reanimation. The science of restoring life!'

Shadows at the edge of her consciousness. She was in a lucid and terrible nightmare.

'It was fortunate that you happened to bring your nephew with you this evening. Children, it turns out, are generally better subjects than adults, presumably because the life force is stronger in them, they react better to the electric spark. Of course, your boy may be too weak constitutionally, and you are probably too old, but I will nevertheless try.'

She was sitting up fully now, remembering Lefèvre's talk of resuscitation, the wand with its spray of sparks. 'You have been killing children and trying to reanimate them.'

'Not merely trying.' He sounded annoyed. 'In one instance I succeeded. He lived only for a few minutes before returning to the other side, but it was nevertheless a success. This is why I must persist, you see, until the answer is within my grasp. And remember that hundreds – perhaps thousands – of children die in this city every year: newborns left to freeze on steps, foundlings starved in orphanages, and so many others steeped in squalor. I have merely selected a few from that great swarm of misery and used them for my research.'

She stared at him. He was insane. He had decided some lives were entirely without value to justify pinching them out. She cast her eyes around the room: vials of liquids, glass bottles of powders, mortars and pestles, books and bowls. Nothing within her reach. She had to keep him talking. She had to keep him away from Émile.

'My mistress worked it out, didn't she?'

He grimaced. 'Unfortunately, yes. Like you, she asked too many questions, peered around too many corners. I thought she might have understood. She was, as you'll recall, an original sort of girl, with a reasonable understanding of medicine. But I suppose a convent education inhibits the mind rather. She ran.'

'So you chased after her and killed her.'

'As I said, unfortunate.'

Unfortunate. Was it anything to do with fortune? It seemed to Madeleine at this moment that it was nothing to do with luck or chance, but with man's ability to justify his own evil, to treat other people as human scrap. She stared again at Émile's prone form. He was so small, so helpless.

'You needn't worry,' Lefèvre said, following her gaze. 'I didn't knock him out – it would have been too dangerous given his weakened state. I told him you'd fainted and then merely put a sleeping draught in his chocolate. He's now quite happily asleep.'

Madeleine's confused mind raced to unscramble everything. Was there anything she could do or say to convince him to release Émile? Any bargaining chip she could use? But she had nothing, had saved nothing, had thrown it all to hell.

'You killed Victor.'

'Victor?'

'The black slave boy.' Lefèvre had not even known his name.

'Ah, yes. I have been trying different types of subjects to see if they respond differently to the treatment. I prefer, by the way, to use the words "put to sleep", as that is how it is done. There was no violence. I assure you almost all of the subjects came voluntarily and were well treated.'

'Until you "put them to sleep".'

'Exactly.' He was fiddling with a contraption made of a wheel and wires, a glass jar full of water attached to it. 'The machine still needs work. Really, I could have done with Reinhart's assistance, but I found he would only go so far with these ideas. He has some old-fashioned beliefs about the sanctity of life, and of course now his own is draining away.'

She remembered what Edme had said when she first came to the house – that her master thought all human life to be precious. Peculiar he might be, but he'd never have helped Lefèvre. He'd have seen him at once for the monster he was. 'Reinhart didn't know what you were doing, did he?'

'No. I was waiting for the right moment at which to discuss the matter with him, but then there was the business with Véronique.' He shook his head and continued with the machine.

The business. The business of running her down with a horse. Madeleine cast her eye about the shelves for something she could use as a weapon: a smashed bottle, a knife. And then she saw – behind Lefèvre – his surgical equipment laid out on a cloth: a scalpel, a lancet, an amputation knife, an array of sharpened steel. But even if she could extricate her hands, there was no way she could free her legs and there was nothing within her grasp.

'It hasn't been you, though, operating the clockwork doll.'

Lefèvre froze for a moment, when he spoke again his voice was tight. 'No. No, that must be some trickster at court who wanted to cause mayhem.'

'And who knows what you did.'

He did not look at her. 'I don't see how they could have done. In any event,' he said more brightly, 'I have spoken again to the King and I think he may be seeing the light. I think the doll will be destroyed and that will be the end of its little games. The focus instead will be my work.'

The King. The King with his love of automata and his obsession with the dead. 'Does the King know?' she asked softly. 'Does he know anything about this?'

'My dear, it is for him that I am pursuing this project. He is my royal patron.' Lefèvre smiled at her as though she were a stupid child. 'It was Louis who wished me to investigate reanimation to the full, once he learnt of my particular interests. A noble ambition on his part, I think, to wish to further scientific research in this way. So you see there would have been little point in you trying to raise the alarm in any event.'

That was what the doll had meant when she wrote her message: that the child thief was protected by the King. It had been hopeless from the first.

'He was concerned, of course, that there should be little trouble, that we should use only the unwanted children of *le bas peuple* as subjects, for their existence was so miserable in any event.'

'So their lives were worth nothing.'

'Sacrifice, Madeleine, is necessary for progress. Every breakthrough has had its sufferers. Every cure has been achieved through pain. Just imagine for a moment that we were to succeed – that we were able to renew life, find the secret of immortality. It would change everything, flood our dark world with light. Think, my girl, of that!'

Madeleine's head was almost clear now, though the pain continued to pulse. Lefèvre was pulling the machine towards her. She had, above all else, to buy time, to keep him talking of his achievements.

'How will you ... put me to sleep?'

'Asphyxiation – a lack of air. You've heard no doubt of criminals occasionally reviving after being hanged. The foolish public think these are miracles, but I see them as something else: as proof that in certain circumstances – where the forces of life and death are balanced – life can return. I have tried drowning also, but it is through asphyxiation that we have had the best results.'

Lack of air. Would he strangle her, then? Would he stop her mouth? She couldn't let herself become numbed with fear, she had to keep her head.

'You didn't try to bring Véronique back to life, though. You left her in the street.'

'Well, yes, those were regrettable circumstances.'

'And yet she *has* come back to life.'

Lefèvre turned to look at her, his expression sour. 'That is not life. That is a clockwork doll. It bears no real resemblance to life. It is nothing like a real girl.'

Only that wasn't true. She *was* very much like the real Véronique – was strung halfway between life and death – and she'd spoken to reveal the truth. 'How is it that the doll speaks, then?'

'You mean that ridiculous song? A trick.'

'No, people say she still speaks and whispers, though she's locked in the room on her own.'

'This is the stuff of peasant folklore.' He gave a short laugh but he had grown very still. 'What do these idiots claim that it says?'

'That she is Véronique reborn, come to take revenge on the man who killed her.' It was an embellishment of the things she'd heard, but also a theory she'd begun to believe.

'This is some vile trick, some joke by an embittered courtier.' He was angry now. She had misjudged. 'No one knew her death was not an accident. No one. And no one will even notice yours.'

He was behind her and she understood that it would happen now. Now, before she was ready. He forced something over her head so that she could neither see, nor fully breathe. A hood, like a hangman's hood, the coarse fabric against her face.

'Please,' she tried to say through the hood, though it clung to her lips. Then, as she felt the cloth tightening, a cord about her neck, she shouted: 'Get off me! Let me live!'

And it came to her as she struggled against him, fighting for breath, that she wanted to cling onto life with her teeth and nails. She wanted to be loved, to love, to cry, laugh, grieve, and she was capable of all those things, she deserved them. She could not let him win. Then she knew that to live, she would have to pretend to die. For a moment she ceased to struggle and let her body go limp. In the instant that she felt Lefèvre loosen his grip, she threw herself backwards so that her head met his face and through her own searing pain she felt the crunch of cartilage, heard his shout of shock. And then came a crashing sound from outside. No, it was *inside* the house. Footsteps growing louder, a man's shout, then the sound of the door crashing open and, from Lefèvre, a strange, almost inhuman howl.

'This is not possible,' he cried. 'Get out! Get away from me!'

Madeleine wrenched the hood from her face and pulled herself up on her elbows so that she could see. There before them was something that could not be there and yet was: the clockwork doll, in her silver dress, no longer sitting at her desk but walking through the doorway and towards them.

28

Lefèvre was backed up against one of the bookcases, blood streaming from his nose, his face very white against the red.

'Who is doing this? Who is taunting me? Because I assure you they will pay.'

Silence, and then the doll began, very slowly, to walk. Her movements were stilted, lopsided, but they brought her ever closer to Lefèvre, who was trembling like a terrified child.

'Keep away from me!' He picked up one of the preparation jars and threw it at the doll, but it missed, crashing instead to the floor, its contents leaking into the rug. 'Who made this? Was it really Reinhart? How in God's name did he do it?'

Madeleine too was shaking, her teeth chattering. Now that the doll was closer to her, Madeleine could see that the green glass eyes had become horribly real, the porcelain given the sheen of skin. Was this what Reinhart had meant when he told her that Véronique lived on? Had he somehow turned his clockwork doll into a living, walking being?

'Talk to me, if you're so clever: did Reinhart do it?' Lefèvre demanded, his voice nearly a shriek. 'Did he raise you up?'

The doll came to a standstill a metre or so away from him. And then came a voice, lower than Véronique's, slower, but true: 'I raised myself, with a little help. Raised myself up from where you ran me down and patched myself back up.'

A hand on her arm. Madeleine almost screamed but she saw that it was Joseph, real and close, gazing at her with concern. With a penknife he began to cut the cords that bound Madeleine's wrists and ankles. 'Don't worry, Madeleine,' he said to her softly. 'We'll get the boy free next.'

Madeleine stared at him. 'I don't understand. What's happening? How can the doll be here?'

Joseph touched her face very gently. 'Look at her, Madeleine. Look at her.'

Madeleine looked, and reality seemed to tilt. Madeleine was back in the great mirrored room at Versailles watching the doll write apparently on its own. She was hearing the voice singing, the voice that she heard now. She was remembering Reinhart's insistence that no one touch the pedestal, no one try to move the doll. She was remembering her mistress lying looking at the ceiling, telling her she'd succeed.

And then she understood. Véronique had been inside the doll. She had been here all along.

29

Véronique

She was close to Lefèvre now, so close she could see the tiny red veins that curled in the whites of his eyes, the specks of stubble on his chin. This man who'd become so obsessed with his own quest for significance that he'd treated others as human flotsam. This man who, when she'd worked it out, had decided to crush her as well. 'Your work is not progress. Your work is an abomination. You've taken the most vulnerable and used and abused them. You'll burn in the pit of hell.'

'I did as the King himself wanted me to do, the King: God's emissary on earth!'

Ah yes, God, who could always be called upon if needed to justify inhumanities. Wasn't it God who'd spoken to Soeur Cécile to tell her to test her pupils? Wasn't it God who must have decreed that Clémentine's soul wasn't strong enough? Not her God. He hadn't done that, nor did he live through the King, whose distinctly human hands had tried to paw at her flesh.

'He's just a man,' Véronique said to Lefèvre, 'same as the rest of you.' Her eyes remained locked with his.

'It is about the progress of humanity, don't you see that?'

Lefèvre's voice was high-pitched, wheedling. 'Certain lives must be cut short in order to advance mankind!'

She stared into his bloodshot eyes. 'Have you really convinced yourself of that? That you can do away with the vulnerable to prolong the lives of the rich?'

For a long moment he was silent. Of course he had no answer. He could only stare, his pupils pinpoints of darkness.

'Mademoiselle, come towards me.'

The voice of her maid, who'd spied on all of them. By rights she should hate her. 'Leave me be.'

'Please, Miss. Step away.' Madeleine was moving towards her, her face covered in blood.

Véronique turned back to see a flash of light above her – a knife. Then a rush of air as a dark shape moved in front of her and there came a shout, a scream, a crashing sound as Lefèvre's body slumped back against the shelves and a slew of glass vials and little ceramic pots shattered on the ground.

For a long moment she stood motionless, watching Lefèvre's still shuddering body, his hand pressed in vain to his chest, the growing pool of blood around him. Véronique was vaguely aware of Joseph moving closer to Madeleine and of the clatter of steel on the floor.

Then came the sound of boots on the ground outside, the shouts of military men. 'Open up! In the name of the King!'

But it was of course not the King who had sent them. It was Madame de Pompadour.

30

Veronique

'I decided, you see, that it was easier to be dead than alive.'

Véronique sat opposite Madeleine in her father's workshop back at the Louvre, among the scattered papers and discarded parts. Outside they could hear the occasional volley of gunshot, the distant beating of drums. From the rooms upstairs, the soft murmur of voices interspersed with an occasional cry; Reinhart had been extricated from his cell but it wasn't yet clear if he'd live.

She watched Madeleine's drained, dirt-streaked face, the face she'd wanted to see for so long, despite everything she now knew. 'And I'd worked out you were beholden to the police, so I had to keep it even from you.'

Madeleine squeezed her eyes shut. The skin around them was raw. 'I feared you'd worked it out,' she said quietly. 'All I can say is that I thought that it was my way out of something worse. My way of saving Émile.' She twisted her hands in her skirt. 'And then, once I regretted it, I couldn't get out. I was trapped.'

Véronique had guessed that when she saw the burn mark, had known Madeleine was being coerced. It had explained so much – about her restraint and watchfulness, her continual holding back. Then she'd found the note, half burnt in the grate. *You are to come to the Châtelet immediately.* Why else would they summon her to the police headquarters? 'You lied to me. To all of us.'

'Yes.'

'You pretended to be my friend.'

'I didn't pretend that.'

She searched Madeleine's face: the beautiful side, and the side someone had destroyed. *Something worse.*

'Your nephew is still asleep?'

'He woke for a bit, confused, but yes, he's sleeping again. Lefèvre clearly dosed him heavily. I suppose I should thank him for that.'

For a while they sat, silent, thinking. Pompadour's men hadn't asked who'd wielded the lancet that wounded the surgeon, merely ordered that they leave at once. Together they had come back to the Louvre, to recover, and to wait.

Madeleine lifted her eyes, shot through with red lines, to hers. Véronique thought she might plead, try to explain herself further, but she didn't. She said instead, 'Will you tell me what happened? Will you tell me how you made the clockwork girl?'

Véronique half smiled. 'Through hard work and imagination; the same way my father's inventions have always been created. Louis wanted Father to make an automaton that truly breathed and bled and moved. Vaucanson had tried and failed so my father took over from where he'd left off, toiling for weeks to make the mechanism that allowed the Messenger to be programmed to write as she did. He worked night and day to make her breathe through bellows, make her cry false tears of water.' She paused. 'I helped him at first, as you'll remember. But then, of course, I died.'

Because, after days of doubt and fear – of piecing together fragments she'd seen and heard – she had finally found her proof: had seen Lefevre carrying from his carriage the prone body of a child. For an instant she was there again, running away from Lefèvre's and towards the lights of the Place Baudoyer, her heels slamming against the road, her heart pounding in her chest, the whir of carriage wheels growing ever louder, the smell of the horse's flesh, and the screaming of her own blood. He had left her face down in the mud, unmoving, and he had made a terrible mistake: this man so obsessed with the line between life and death had wrongly assumed that the girl he'd run down was dead.

'Your father knew then, right from the start, that you were alive? But a man came to the Louvre with a note. Reinhart said he'd identified your body.'

'The man brought a note from me, telling my father to come to the home of the woman who had found me and was looking after me.'

He had arrived in the woman's poor little house confused and disbelieving. Véronique hadn't anticipated how distressed her father would be, almost beyond words, when he saw her broken state; when she told him of what had happened. 'My daughter,' he'd kept saying, 'my daughter,' as though only realising now that she was his. 'I have failed to protect you,' he told her. 'I was so preoccupied with achieving what the King had asked me to do that I failed to notice what he himself was doing and supporting. I have been lost in dreams of moving machines, when I should have been focused on you.'

The horse's hooves had largely missed her, but the damage was still severe. After a time, he moved Véronique back to their old house on the Place Dauphine, but kept the shutters barred and the steps unswept to give the impression that the house was still vacant. He arranged for a nurse to care for her, a bent-backed beetly old woman, the same woman who would later lay the body out for Véronique's funeral; the wax mask her father had prepared long before laid on the anatomical doll. A horrible trick, yes, but they needed everyone to believe she was truly dead. It was the only way she would survive.

At the Place Dauphine, she lay recovering, thinking. Reinhart visited Véronique whenever he could, sitting for long stretches at her bedside, silently chewing his nails. 'They must be stopped,' he would occasionally say. 'They must be punished. For everything they have done.'

She had told him then, not just of Lefèvre, but of what had happened at the convent: of how she'd tried to help Clémentine after they locked her up – sneaking her food at night-time to the darkened cell where she'd been kept for days; of how she'd failed to recognise just how ill the girl had grown, despite her listless manner, despite her twig-thin wrists. Of the terrible shame that had burnt through to her bones when one night she ran to the squalid cell and found her cold and dead.

'You should have written to me,' her father had said. 'You should have called on me at once!'

'But you were far away, Father, and I barely knew you.'

She had been fifteen years old and frightened. She'd had no idea what he would say. She'd not anticipated in fact how she herself would react – consumed not just with grief but with a burning, blistering fury.

In their old house in the Place Dauphine, Véronique had spent three days lying in her bedroom recovering, thinking, listening to the growing sounds of violence in the streets outside. Every day when her father came he would tell her of what progress he was making with the clockwork doll, and of the growing unrest, and they would talk about how they would leave Paris, and whether there was some authority to whom they could go. But of course there wasn't, because how could you challenge the King? Who on earth would believe them? And how would you be able to trust whoever you told – the police, the commissioner, some government minister – not to run straight to Louis himself?

On the third day, as grey dawn seeped into her room, it came to Véronique at last: they would take the creature the King himself had commissioned, and they would adapt and twist her against him. They would take the nearly-made clockwork doll and employ her to expose the truth.

Her father had been hesitant at first – it was too dangerous. But what was the alternative, Véronique demanded? For them to run away from Paris and pretend she'd never seen what she had? To let Lefèvre continue to take children from the streets for his own insane experiments? How many more of them would there be? How many had there been already?

And yes, it was in part that, the desire to stop them, the desire to save the children in a way she had failed to save Clémentine. But it was also a desire to hurt, a desire to haunt. These men who'd always had exactly what they wanted, ordering that children be scooped off the streets as though they were sweets in a shop window – she wanted them to suffer. She wanted them to learn. Louis had demanded something astonishing and, by God, they would deliver it.

'And you were hidden within the pedestal,' Madeleine said.

'At times, yes. There are, it turns out, some advantages to being small.' The hiding place had been her idea, her plan. *'We'll put the doll on a pedestal, Father, as Vaucanson did the Flute Player. And then I can climb inside.'*

'But they said scientists had inspected the doll.'

'Yes, they did, and I was careful to hide elsewhere when that happened. I was only inside the doll when we needed her to sing or whisper, or on the occasions we needed her to move on her own. She could write and cry all by herself, providing we programmed her in advance. But that last time in the *Galerie des Glaces*, when the Messenger gave her first real clue, the time we put cochineal in the water so that she'd seem to cry tears of blood – I was hidden within. It was I who sang, I who made her write the second message, telling of my own death.'

She'd stayed inside the pedestal as they rolled the doll away and locked it up. Her chest had been so tight she could barely breathe, with fear and exhilaration. She'd stayed in that room for two days, existing on the food and water she'd brought with her – two days of near madness when she'd known that it was all on her now: to make them all believe the Messenger was real. To make them believe everything it said. At night she had sung the folk song Clémentine had often sung:

> 'Lend me your quill
> To write a word.
> My candle is dead,
> I have no light left.
> Open your door for me
> For the love of God.'

She had remembered that evening two years ago, on the day Clémentine had died: the nuns and novices standing in chapel, assembled to pray and sing on behalf of their fallen sister. Some were muted, some were weeping, many ashen with fright, others with guilt and horror. And as Soeur Cécile came forward to speak, one of the girls let out a piercing scream. For behind Soeur Cécile, the statue of Jesus was weeping. But he was not crying normal tears of woe. He was crying tears of blood, tears that streamed down his tortured body. So that – whatever Soeur Cécile might try to tell herself, however she might attempt to justify what she had done – she would know that God was punishing her, and that her soul was forever damned.

No one had known Véronique was behind it. Who would suspect a quiet young girl of planning such a revenge? Had the nuns paid better attention to the books from which she studied – had they seen the black-tongued devil and the mechanical Christs – they might have had pause for thought. But only Clémentine had known of those, and Clémentine had been sapped of life. That secret knowledge was the one thing that had kept Véronique going, kept her alive, until the carriage had come to take her to Paris, away from the place that had killed her only friend.

In the same way, no one had guessed that it was her making the doll write and sing – not the scientists, the spectators, the servants, the King. Not until Madame de Pompadour had paid a visit and outsmarted all the rest.

'You told her everything?' Madeleine said.

'She guessed half of it on her own. She said she would see to Lefèvre – she'd come to Paris to oversee it herself. That's why I ran to her when I learnt from Joseph that you'd gone to him with your nephew.'

'I thought him kind. I thought he'd helped. Not much of a spy, after all.'

'He'd perfected the veneer of a decent, kind man to conceal the void inside. You weren't to know that. None of us were. Even Father was taken in, and he'd known him, remember, for years.' His only true friend and it turned out he was as false as the paste in a harlot's brooch.

'What about Louis, though? Who's to say he won't just employ someone else to continue Lefèvre's work?'

'Pompadour says she has it all in hand, but I don't know how much power she has.'

'I believe she controls the police, or at least they recruited me on her orders.' Madeleine averted her eyes. Her face, Véronique noted, was greyish white with misery and shock.

'Madeleine, you said earlier that you were trying to escape. Escape what, exactly?'

She saw Madeleine stiffen. 'You don't want to know, Miss.'

'You think you can shock me, after everything I've seen these past weeks?'

'Perhaps.'

'Go on, then.'

A pause. 'I told you once that I'd had a little sister who died.'

'I remember. Suzette.'

Madeleine nodded. 'Émile's mother. She died in childbirth. She didn't have a proper surgeon or anyone like your father because my mother wouldn't pay for one, only some woman who claimed to be a midwife. And the baby wouldn't come out. Two days it took and then she died, with the baby still inside of her. She was nineteen years old.'

'Nineteen. But that means that Émile—'

'Was born when she was twelve. Yes.'

Véronique sat back down. 'Who is his father?'

Madeleine's face twisted into a warped smile. 'The father could have been any one of many men who were paying our mother to lie with her.' She nodded and Véronique felt the skin on her own face freeze. 'So I grew up hard. I had to. And then, when Suzette died, something in me died too. I don't say that as an excuse for what I then agreed to do, but that's how it was. I didn't care about you or any of your kind. I didn't really care about anyone. I only wanted to get out. To get Émile out. That was what I'd promised Suzette.'

'And the police paid you well, of course.'

Madeleine snorted. 'Oh, they *said* they would pay me. I had a stupid dream I would use the money to set up a *oiseleur*'s shop like my father's, keep Émile with me. But they never paid me, not a sou. There was always something else I had to do or they would punish me, imprison me, take me away from Émile.

And now they'll say I killed an aristo. So you see it was all for nothing.'

Véronique stared at Madeleine. Could she blame her for what she'd done? She'd have done the same herself, perhaps, if her prison had been not a convent, but a brothel.

'And now?'

Madeleine shrugged. 'I could try to run, but there'd be little point. They'll have orders to stop me at the city gates, and the police have taken my papers.'

'You assume they will punish you, but no one knows what happened inside that room apart from you, me, Joseph. And so far, no one has asked.'

The door opened and they both turned to see a small, ruffled-looking boy, rubbing his face.

'Madou?' he said groggily. 'Where are we? Why are we here?'

Madeleine ran to him and hugged him. 'We're at my mistress's house, *mon petit*. You're safe.'

The boy looked about him – at Véronique, at the room – suspiciously. 'But where is that man, who gave me chocolate? He said I could have more today.'

'He's gone now,' Madeleine told him softly, 'but Edme's chocolate is the best you ever tasted. Here, come with me.'

Once in the doorway, Madeleine looked back at Véronique and for a long moment they stared at one another. Then she led Émile out of the room, holding him by the hand.

Jeanne

Lights had been strung up across the palace gardens so that the entire place twinkled like the night sky above. Coloured silks had been draped across the clearing and incense perfumed the air. Alexandrine would adore this, Jeanne thought, looking about her through her mask. She would bring her back to visit this very week. Louis couldn't refuse her what she wanted, not now that she knew what she knew.

Berryer sat on a stool to her right, close enough that they could speak confidentially. The strains of violins reached them from across the lake. They were, to any passing courtier, simply a genteel couple enjoying the masked ball and each other's company.

Close up, however, she could see the veins on Berryer's temples, the tightened muscles of his jaw. 'But how can you be sure, Marquise, that Louis knew of all of this?'

'Because it fits. It fits with everything I know about Louis, everything I'd feared. And because, frankly, Nicolas, why else would the clockmaker and his daughter have done what they did? Why else play that outlandish game? If they'd thought only the surgeon was involved, they could have gone straight to you.'

She'd been terrified at first, hearing the doll whisper in that shadowy room, telling her that Lefèvre had taken the children, telling her that Lefèvre had killed Véronique, and that Louis knew all about it. For a few moments she had truly believed it – that this doll could really speak. But then she had heard the waver in the doll's voice, had realised that the sound came not from the automaton itself, but from close by. She'd pulled back the velvet curtain and in an instant the magic and the horror had disintegrated. For there stood a girl – a slip of a thing – arms wrapped about herself. Jeanne had brought her lamp closer to the girl's face and almost gasped, almost laughed, all at once.

'You! You, of all people.'

'I had to stop them.'

She'd examined the girl's expression in the light. Was she insane? She was certainly filthy, her hair plastered across her forehead. She must, Jeanne realised, have been in here for days, ever since the doll had been sequestered in this room. 'Come with me,' she'd whispered. 'Quickly and quietly. You will tell me everything. You will tell me what you know. And then I will decide what to do.'

Behind his mask, Berryer looked grey, she thought, as though someone had leeched him entirely of blood. 'There is a letter. About certain amounts paid over. I will show it to you. I had hoped it was a bluff, but perhaps not. I just cannot believe that the King would be so … so …'

'Stupid? It seems staggering, doesn't it, that a man in his position could have failed to consider the consequences of such a project, or at least been blind to the full extent of it – but I suppose he convinced himself he was advancing the cause of humanity.'

'If it were to come out, there would be chaos, outrage.'

'Oh, yes. What we've seen so far would seem merely a ripple.

There would be full-scale riots, there would be revolution.' She would lose her place, and he his kingdom. 'But it will *not* come out. It can be blamed entirely on the surgeon who, being dead, will be unable to protest. You still have his body?'

'Yes.'

'Is it known how he died?'

Berryer shook his head. 'We haven't yet questioned those in the room and arguably there is no need.'

'I would agree.' She wondered if he knew who had struck the blow. She had formed her own opinion.

'It may be best, in fact, to say the man was killed by the guards to prevent him from further outrages. And we could perhaps arrange a public spectacle of some kind. A beheading or some such. The people like such things.'

'That would, I think, be cathartic.' For herself as much as the crowds. The arrogance of the vile, pompous little man – to think he could take people's children and experiment upon them as though they were less than dogs. To guide Louis into such squalid evil and pretend that it was science.

There came a gust of laughter from a group of people not far away, watching an acting troupe.

'You haven't confronted him directly?' Berryer ventured.

'No.' She had not, in fact, been able to bring herself to speak to Louis at all. She'd been feigning illness since the previous night and had kept her apartments locked. He would grow angry with this soon and demand that she came to dinner as he often did when she pleaded a headache, but she found she didn't much care anymore. His anger was nothing compared to hers, and he no longer held all the cards.

'I will speak to him tomorrow. In the meantime, make sure that the clockmaker and his daughter, the maid too, are left alone.'

'But they know too much.'

'That is why I have this very day offered them a bargain: their silence for their safety and a guarantee the King will be controlled.' Véronique had come to see her that morning. She had been adamant: they would only agree not to speak publicly if the families of the children were compensated, her maid properly paid and protected, and the King's proclivities curtailed. A dangerous request for the girl to make, and she must know the possible consequences. The girl was certainly bold, and Jeanne found that she rather liked her for it. She guessed too that Véronique sought protection for her maid because it was the maid who had killed Lefèvre; a woman who'd been scarred by a man in the past would be quick to take up a knife.

'It would be simpler ...' Berryer spread out his hands.

'To dispose of them all? Yes, it probably would, Nicolas, but given that it was the girl and her father who brought the whole thing to my attention, I find I am not minded to do that.'

Berryer considered this. 'How will we explain the girl's apparent return to life?'

'A medical miracle, perhaps?' she said lightly. 'Louis surely would approve of that. Or we send them away, outside Paris, or abroad. That is to be decided, but for now, don't touch them. Your priority is to find a way of presenting this to the public that will calm them down and end the riots. They will need to feel the matter has been properly investigated and that justice has been done, et cetera.'

'Only it has not.'

She raised an eyebrow. 'Not exactly, no. But then that has never bothered you before. Versailles has its own system of justice. Leave me to speak with Louis. After this, I think additional levels of scrutiny.'

'Constant, Marquise. Constant.'

She nodded. She had turned aside from Louis' forays into

the city, dressed as a duke; she'd indulged him his morbid, gruesome lessons. But there must always be someone watching now. His cage must be complete. She felt a pinpoint of sadness, but it fell to her to protect him. She did so out of love, or at least the shadow of it that remained.

'Have you found anything, by the way, to confirm who has been writing those nasty little verses accusing me of using the dark arts for my own nefarious ends?'

'Not as yet, I'm afraid. They have been careful, I fear.'

'Perhaps. But we know who it is, and it doesn't much matter how careful they are now. Louis will have to punish them, irrespective.' Because she would let him know – subtly, gently – that if he didn't, she could act. The balance of power had shifted. 'And you must pay the *mouche*. The maid. What you owe her, and more.' If it was indeed she who'd got rid of the man, she was worth her weight in gold. 'I can't understand how she hadn't already been compensated.'

A twinge in Berryer's jaw. 'I think you overestimate her worth.'

'Meaning what, exactly?' There was something there. 'Ah, so you knew her yourself.'

'It was merely that my junior officer had some concerns.'

'If by junior officer, you mean that oily young man with differently coloured eyes whom I saw at the Châtelet then you should know better than to listen to such a grubby little person.' She'd recognised his type: power-hungry, spiteful, sadistic. Similar indeed to Berryer, but without the superficial charm and cunning that made the latter worth keeping in his place – or at least it had until now.

'Perhaps, but he has his uses. It was he who prised Reinhart's original maid out of the house and he who placed the *mouche* within it.'

Jeanne's eyes narrowed. 'And how did he remove the

original maid from the house?' She could guess, of course, but she wanted it confirmed.

Berryer looked as if he'd swallowed something bitter. 'She fell pregnant.'

'Well, that is an unfortunate way to fall. Has she been paid off?'

'I do not believe it was considered necessary.'

Jeanne regarded Berryer closely. No, perhaps he was not sufficiently different to his underling to justify her continuing support. 'Nicolas, when I said I wanted the clockmaker vetted, I did not say I wanted domestics ruined without pay, did I?'

'There are always consequences, Marquise. In any spying operation, there are always inconvenient people who must be removed.'

She exhaled. 'Where is the girl now?'

'I've no idea.'

'Find out. Compensate her, if she's still alive. And excise that carbuncle of an officer of yours. I don't like him. I don't trust him.'

'Yes, Marquise.'

'Good.' She sat back and listened for a moment to the music, the violins soaring. 'We will need a new clockmaker. I will try to find someone unspeakably dull. Without attractive daughters.'

★

Jeanne heard the sound of his footsteps but continued her work with the tiny chisel. It required all of her concentration. The gem carving was nearly complete.

'Where on earth have you been these past days?'

She did not turn. 'I have been here, Louis. Working.'

'You've not come down for dinner, not appeared in the

petits apartments. Are you sick again?' There was no sympathy in his voice, only cold anger.

Oh yes, she was sick. The consumption that had brightened her eyes and flushed her cheeks was destroying her from inside. But she would not tell him that. It would be the end of her. Jeanne squinted at the carnelian through the magnifying glass. Yes, it was almost perfect: a portrait of Alexandrine etched into the shining surface. She would give it to her that very afternoon, for she had sent a carriage to collect her.

'Well?' he said. 'Answer me!'

Finally she turned to him, the chisel still in her hand. 'Did you know, Louis,' she said conversationally, 'that your surgeon, Lefèvre, has died?'

'What?'

'Yes, in his workshop.'

'How?' His fleshy face had gone quite white.

'There was an incident. It was found, you see, that he'd been carrying out certain experiments.'

Louis remained motionless, but his eyes darted about.

'What did you know of this?' she asked, keeping her voice low.

'What do you mean, "what did I know of this"?' he blustered. 'I knew he was a scientist, a progressive. I didn't involve myself with his experiments.'

'You didn't? That's not what he claimed.'

He was sweating now. She could see the beads of perspiration standing out on his forehead. 'Then the man, *mon tresor*, was a liar.'

She nodded and turned back to the small worktable to continue on the gem.

'From whom did you hear this nonsense?' he said sharply.

'From Véronique, the clockmaker's daughter.' She suppressed a little smile.

337

'But she's dead.'

'No, it turns out she is alive.'

His eyes were round, the whites shot with blood. 'That isn't possible.'

'Now, Louis, you yourself have always said that reanimation must be achievable, haven't you? Well, now she is your proof.'

'I don't believe it.' His voice was strained. 'You are toying with me. Where is she?'

Jeanne walked over to her writing table where Miette was chewing some paper, and pulled out the letter that Berryer had handed her last night. 'Officers found this in Lefèvre's office. In it, he writes of his experiments and of how you were his patron. He lists the occasions on which you gave him money and the sums, which can be checked, I would imagine, against the royal accounts. I suppose he wrote it as a sort of insurance policy in case the whole thing came to light. Fortunately, the officers who attended his house handed it straight to Berryer, who passed it on to me.' She held the letter out to Louis but he did not take it. 'So I will ask you again: what did you know of the experiments? Do be honest with me, *mon cher*. You know I can tell when you're lying.'

'You would not understand.'

She raised her eyebrows. 'Try me.'

He coughed and began to pace the room. 'Lefèvre believed he could perfect the art of reanimation, that he could discover the secret to life and its renewal. He believed he was very close. Can you think of anything more important, Jeanne, anything more vital than finding a means of restoring the vital spark of life?'

'Why, Louis, did this necessitate the killing of children?'

'No, no, you misunderstand. Lefèvre merely found children who were close to death and removed the final life force from them.' She could see the Adam's apple move in his throat as he swallowed. 'He assured me that they would not have survived long in any event, and that they all died in their sleep.'

'Even if that were true, why take live children at all? Why not use corpses, Louis? One of the many dragged from the Seine?' Jeanne thought she knew the answer to the question, but she wanted to be sure of what he knew.

'He'd tried that, tried all sorts of things. Lefèvre said that the only way to obtain any real results was to use a subject freshly dead, and the one success he'd had to date was on a young boy.' He pulled at his collar.

'So you agreed that he should find further young subjects.' Still she spoke quietly, calmly.

'Only ones close to death, I tell you! And I insisted, I'll have you know, that he only took those who were likely to die and unlikely to be missed in any event – street children, foundlings. But he found that their malnourished and weak state meant he had no further successes.'

'And thus he began taking the children of tradesmen.'

'I did not ... we did not discuss that.'

'But you knew that was what was happening – you must have – because it was being reported. Because people have been rioting in Paris. You must have known he was taking more children, perfectly healthy children. And yet even then you did not stop him.'

'How am I to know whether what is reported is true? Most of what they write is lies! He believed he was close. Very close. Imagine for a moment, Jeanne, just imagine—'

'No! *You* imagine,' she hissed. 'You imagine what would have happened if your people had realised that you were in-volved in this, that you had effectively sanctioned the killing of their children. Do you think they would understand? Do you think they would see you as a great scientist? These are the people who've been claiming you bathe in their blood.'

Louis looked petulant now, his bottom lip puffed out like that of a sulking child. 'But they do not know,' he said stiffly, crossing his arms.

'No, *grâce à dieu*, they do not know. And I will do what I can to ensure this is smoothed over, and that Lefèvre alone is held up as the villain.' She exhaled. What did that make her? Complicit in their deaths? 'But there must be no more "lessons", Louis, no more experiments.' She walked over to the window and looked out over the forests of Marly. 'We will bring a new scientist to court, yes? Someone entertaining.' Someone anodyne who did not deal in human flesh. She folded Lefèvre's letter and secreted it in her corset. She would keep this and make a copy. Her own form of insurance.

After a long moment, she sensed him moving towards her. 'You are always good to me, Jeanne.' She felt his hand at the nape of her neck and flinched.

She moved away. 'I have been meaning to speak to you about certain other matters, Louis.'

When she turned, he looked so dejected that for a moment she felt sorry for him and the unnatural life into which he'd been born. But he was a grown man now – a king. His decisions were his own.

'What other matters?'

'The verses have continued to circulate about me. Those nasty little *poissonades*. They distress me, as you know, and they lower my reputation. They have become increasingly dangerous.'

He stared at her but said nothing.

'You and I both know who is behind those verses. You and I both know who it is who poisons others at court against me.'

'You have no proof.'

'No, because he's clever and because he has others working for him. But I don't think that matters anymore. I want you to send him from court at once.'

She saw Louis' jaw tighten. 'I cannot. Not without evidence.'

'My darling, you are the King. You can do anything you like. And I am asking you to send Richelieu away.'

'But I need him as an advisor, and a friend.'

You need not to have me as an enemy, Jeanne thought. She smiled thinly. 'Well, I've been thinking about that, and about my own role. About how it should change.'

'Change? You do not wish to leave?'

'No, no. I wish to remain here as your friend and advisor.'

'My *friend*.'

'I know your desires, *mon ami*. I cannot fulfil them. But I can find others – younger, sweeter, more pliable – who will.' She had thought it all through. It was distasteful, yes, but it was better than the alternative, which was that he'd find his own sources: inappropriate, ambitious, a threat.

His mouth parted and she put her finger to his lips to stop him from speaking. 'All this will leave me with more energy to do the thing that I am best at: looking after your interests, just as I have looked after them today. This will be a new stage in our relationship, the most important stage of all.'

After a long moment, he nodded slowly, understanding this was part offer, part warning, and she smiled at him, wondering at how foolish she'd been to think that the only way of retaining her position at court was to bow to his desires, to become what he wanted her to be. 'So you see, you can do without Richelieu. You will have me.'

'My love, you assume that by asking him to leave court his power will evaporate. It will not. Better to keep him at Versailles where we can see him.'

There might be some truth in that, she thought, but she loathed the man, wanted him gone. 'At the very least you will convey to him in the strongest terms that he will lose his position, perhaps his head, if he persists in spreading rumours about me. I will not be undermined any longer, not by him, not by any of them.'

A half-smile. 'You have thought all this through.'

'Did you expect any less? Is it not my role to continue to challenge you? To continue to protect you?'

He raised his hand to stroke her face and she felt for an instant a surge of the love she'd once had for him, such a powerful and radiant thing, decayed now to a painful mistrust.

'I am fortunate to have a mistress such as you, Jeanne.'

She took his hands. 'A friend, Louis. An advisor.'

'And you will deal with all this? Make sure it goes away?'

'It will be as though it never happened.'

He looked only momentarily uncomfortable and then his attention shifted. 'The doll, the clockwork doll. You were right. It ought to be destroyed.'

'Except, *mon cheri,* that it has vanished.'

'Has it?'

'Yes. You may check the room yourself. She is nowhere to be found.'

She watched him as he left the room, his cape flowing behind him. And then, a moment later, a shadow.

'Good,' she said to Miette. 'Berryer acted quickly.'

Louis would never now be alone.

32

Madeleine

Two days had passed since Véronique returned to life. Still the police didn't batter at the door, nor summon Madeleine to the Châtelet. Instead, she tended to Doctor Reinhart, who improved slowly, and read and played with Émile. The apartments had become a kind of stronghold, a sanctuary from the riots outside. They ate their way through Edme's jams and preserves, dried beans and lentils and peas. The calico cat brought the occasional mouse, as her contribution to the feast. Véronique made up a tincture for Émile's throat following the instructions her father gave her. It wouldn't last, Madeleine knew that all too well, but she'd savour every moment they had.

She couldn't say exactly why her mistress hadn't outed her to the others. Certainly it wasn't that the girl was soft. No, Véronique was no softer than she was; beneath it all, she was silver and steel. Perhaps it was that on some level Véronique understood: she knew what it was to want something out of life, she knew what it was to feel trapped. Perhaps she recognised that the Madeleine who'd agreed to act as spy was not the same Madeleine that appeared now, pale before the drawing-room glass, the ligature burn that ran round her neck now reduced to a thread of red. She touched her hand to it, felt the miracle of the mending skin.

Madeleine saw a shape flicker in the mirror and turned. Joseph stood in the doorway, watching her. 'Does it still hurt you?'

Staring at him, her hand still at her throat, she knew then that she couldn't continue to lie to him, not when she had so little time left. 'I need to tell you something, Joseph.'

He didn't reply, merely stared at her, his expression giving nothing away.

'You'll hate me and I won't blame you, but it's right, I think, that you know.'

He folded his arms. 'Madeleine, if you mean that you came to Master Reinhart as a spy, I've known that for a long time.'

Her eyes widened. 'Véronique told you?'

Joseph shook his head. 'I heard that man bullying you weeks ago, burning your arm, telling you he'd throw you in gaol. I wanted to kill him.'

'Why didn't you report me then? Why didn't you tell the master?'

'Because I knew what would happen to you. And because I saw that you had no choice.'

It would have been easy to agree to that, but it wasn't entirely true. 'I did have a choice at the beginning, Joseph. I made the decision to accept the role, for the money, for Émile. No one forced me. Not really. I should never've agreed to do it.'

'Then you would never have come to Doctor Reinhart's house.'

'No.' She would still be in the Rue Thévenot, her heart closed, dying a little each day.

'And then we would never have met.'

She looked up at him. Still that unreadable expression on his face.

'Would that have been a good thing?'

He didn't answer that. Instead he walked forward and suddenly his lips were on hers, warm and dry, and her head was full of his scent – of lavender water and powder and his own smell beneath. She pressed her mouth back against his. It was the first time she'd ever kissed a man because she wanted to. His arms were around her and she felt the heat surge up through her, the blood warm in her cheeks. He pulled back then, putting his hands on her shoulders, looking into her face, and his smile was like a breath of air.

*

Maman came that afternoon, as Madeleine had known she eventually would. Her skin looked ever greyer, Madeleine thought, the bags beneath her eyes more prominent, the lids whiter, and her breath rattled from the climb up the Louvre stairs as though she was decaying from within.

'What are you doing here, Maman? This is my place of work.'

'Émile still hasn't come back. I wanted to check he was with you and you've not replied to my letter.' Her eyes darted around, pricing up the furniture, the wall hangings.

'He's here, safe, just as you knew he was.'

'Well, then, that's all well and good,' her mother said shortly, pushing her way further into the hallway and whispering at her hoarsely. 'Now, we need to talk about the money. I've been trying to get to Camille, but he's gone underground. His house is all boarded up.'

Was it? That was interesting.

'So you must go to his superiors,' Maman continued. 'You must tell them that he promised you.'

'Of course. Of course that's why you're really here: the

money. There's no point in me going to his superiors, Maman. The police'll say I failed because I didn't work out what Reinhart was about. I doubt I've very much time before they turn up here themselves. I should've demanded the money up front, before I went to the clockmaker's house.'

Her mother became animated at that. 'And how was you to work out the man was building a mechanical devil? Let me speak to his betters, then. They must see—'

'Maman, there's no point. I won't get a sou, and even if I did, you'd never see the shine of it.'

'Now there's gratitude for you!'

Rage flowed through Madeleine's veins then, white hot. 'Gratitude for what? For selling me at twelve, for shutting your ears to your daughters' cries, for leaving Suzette to die, for leaving me in the hands of a man who did this – this – to my face?' She thrust her face closer to her mother's and pulled back her hair so that she could see the full extent of the mark that ran down the right side, the brand of a burning poker.

Her mother averted her gaze. 'I never knew he'd do that, Madou.' Her voice was quiet. 'And as you'll remember, I made the officer leave, never to return.'

'Yes. Made him pay, though, didn't you? For spoiling your goods?'

'It was only right that he compensated us.'

'Compensated *you*. How much was it, Maman? Remind me: what was the price of my face?'

She saw clearly her reflection in the looking glass as she'd unravelled the bandage and seen for the first time the red gash in her once perfect skin. She'd known, without any doctor having to tell her, that it would be there for her whole life – a reminder of the time a man'd been let loose to damage her however he pleased.

Her mother was turning to the door now. 'I don't see why

we need to go over that again, Madeleine. It was a long time ago.'

'Yes, you're right. I was fifteen. Fifteen years old. And instead of going to complain to those higher up in the police, you did your best to cosy up to them. So that the man who did this to me wasn't reprimanded, wasn't punished in any way. So that the man who did this to me is now Lieutenant General of Police.'

'You think my making a fuss would have made any difference to that man's rise? You think they'd have given a damn that he'd beaten a whore? You're dreaming.'

'Maybe, yes. But to say nothing? To pretend it was no issue? Your own daughter?'

'Like I said, I made him pay.'

'And what became of that money, Maman? That money you got for my face?'

'It was spent long ago on your upkeep – on keeping you all in stockings and bread and candles.'

'I don't believe that. I believe it's locked away in that little stash you keep hidden away, and which you count every night, which you could've spent on a doctor for Suzette, but didn't. Which you could've spent on a doctor for Émile. So don't you dare come here asking me for money. You keep away from this house.'

Maman turned to face her. She looked ill, her sagging skin the colour of a grub. 'I don't know why you've turned like this, Madeleine.'

'Don't you? Maybe I've finally woken up. Maybe I've realised I deserve something better in this life, not just for Émile but for me; that I'm not the piece of tat you've sold me for all those years ago.'

Joseph had appeared in the hallway, drawn no doubt by their voices. 'Is everything all right, Miss Madeleine?' He was staring, face rigid, at her mother.

'Yes, Joseph, thank you.' She collected herself. 'Everything is fine. My mother here was just leaving.'

Joseph walked forward, straight-backed as ever, and opened the front door. When Maman failed immediately to move, he said, 'Madame, let me show you out.'

33

Madeleine

She was making a nest for the mechanical animals, packing them in feathers and cotton to keep them as safe as she could: the bat, the mouse, the mechanical peacock, the collection of automaton boxes, all wrapped up and nestled snugly together in a maroon leather case. Véronique and Joseph were dismantling the clocks and wrapping the parts in fabric. A few would join them on their voyage, but most would stay in Paris, shrouded in storage, awaiting their return.

'Not that one!' Doctor Reinhart entered the room, surprisingly quick on his wooden crutches, and pulled away the material from a golden carriage clock. 'This we are taking with us. I want to show him the design. And you must bring your part-made dolls and plans, Véronique. You haven't yet packed them all up.'

'I will, Father,' Véronique said, half amused, half annoyed. 'We're working as fast as we can.'

Tomorrow they were to travel to Prussia, to the kingdom of Frederick the Great, a man so obsessed with automata that when he'd learnt of the clockwork girl he'd demanded Reinhart travel there at once to construct an entire army of marching mechanical men. Madeleine didn't know exactly how this had come about, but she suspected Pompadour had oiled the wheels of the machinery that was to take them so quickly and conveniently far away from Versailles.

Madeleine was nervous, of course, jittery as the day she'd come to the clockmaker's. She'd never left Paris, never mind France itself. But it was a fluttering sort of exhilaration, like the bird's silver wings, not the desperate hope she'd felt when she first left her mother's, nor the creeping dread that'd thickened her blood in the weeks before Lefèvre's death. Véronique would be there, and Joseph. And, thank the Lord, Émile, the useful little machine having become kitchen boy and chief taster of Edme's jams.

'Very well, then,' Reinhart said, 'I will collect up the things in the workshop, and Madeleine, I will need your help.'

'Yes, Master.'

'Father, *pour l'amour de Dieu*, let her finish here. I will come and help you shortly. We have time. Stop fretting. Let us get on with our work.'

'Yes, yes.' He waved his hand at her. 'The old man is making a fuss again.'

Once his back was turned, Véronique shook her head at Madeleine and they smiled at each other. Reinhart had been broken and put back together slightly differently, so that a softer creature sometimes showed itself through the metal carapace he'd constructed for himself.

'There was something you needed to do, wasn't there, Madeleine?'

Madeleine closed the leather case. 'Yes, if it's all right, I should take Émile. We won't be very long.'

Véronique nodded. 'And then you must help me finalise which clothes to take.'

Joseph walked after Madeleine as she left the room. 'Do you

want me to come with you?' he asked once they were in the hall.

'No, thank you, you're needed here. And it's right that we go alone.'

He gently touched her face, the scarred side, and she pressed her cheek against his hand.

Madeleine found Émile, as she suspected she would, standing on a chair in the kitchen and licking a wooden spoon as Edme poured the muffin mixture into the waiting tins. For a moment she stood in the door, watching them. 'How is your new assistant getting along, Edme?'

Edme glanced at his sticky face. 'Still too thin,' she grumbled. 'I begin to think he's got the worm.'

Madeleine smiled, though she was remembering another little boy eating jam tarts, a boy who'd known little sweetness at all. She held out her hand. 'Émile, we need to buy some flowers and say goodbye. Come, let's get you cleaned up. Look at the state of your face!'

Émile clambered down from the chair and came over to her but refused to relinquish the spoon.

'You told me she liked roses.'

'Yes, I did. We'll buy some on the way.'

They walked hand in hand to the Cimetière des Innocents, Madeleine taking out a perfumed handkerchief to try to blot out the stench of death. They didn't know exactly where Suzette was, of course, only that she'd been laid somewhere in the midst of the pauper's pit. They had a place for her now, though; Madeleine had used some of her money from the police to have a gravestone erected. Nothing grand, just a small headstone with the inscription '*Soeur et mère bien-aimée*' – some way of recording her life amid the relentless march of time.

Madeleine placed some of the roses they'd bought at the market at the foot of the stone. 'There. That's something, isn't it? Let her know we're thinking of her till we come back.'

Émile traced the letters on the stone with his finger. 'We *will* come back here, won't we?'

'I think so, yes, Émile. To set up our shop, remember?' She had enough money now: to set up the *oiseleur*'s store of her own. It seemed a long way in the distance, both in space and time, but not altogether impossible.

'With parrots?'

'Yes, with parrots. Why not?'

'Talking parrots?'

'Well, you'd have to be the one to teach them.'

They walked then to the larger stone that'd been placed on the other side of the cemetery, far back against a wall. By this stone fresh flowers had already been laid along with a paper windmill and a framed picture of a solemn, dark-eyed boy. The stone bore no inscription, but everyone knew what it was: a monument to the stolen children of Paris; the boys and girls who Lefèvre had used to conduct his deranged experiments. Some of their names were known but there were, she suspected, many more who had never even been missed: the girl in the doorway in the Rue Thévenot, the rabbit boy from the market. She'd never know how exactly they perished, nor how many more had gone.

Madeleine placed the rest of her flowers at the foot of the stone and stood back to regard it. Louis' role in the whole affair had been carefully concealed. It'd all been blamed on Lefèvre, whose corpse had been beheaded in the Place de Grève as a sop to the anger of *le peuple*. When the anger refused to burn out, three of the rioters had been hanged, one a boy of only sixteen. The riots had died down after that, the debris swept up, the reports suppressed, but the rage continued beneath the surface, simmering, ready to bubble over again when the time was right for rebellion.

Véronique had been furious when she realised there was to be no real punishment for the King, despite his having turned a blind eye to the scale of the killing, despite his funding Lefèvre's supposed 'research'. Madeleine, though, was unsurprised. Justice rarely fell on the heads of anyone above a baker – it was hardly going to come for the King. Since when had the lives of people like these children and their parents been valued by the powerful of Paris or by those who walked the glittering halls of Versailles? That was why, she'd come to realise, you had to put a value on yourself – couldn't trust anyone to do it for you, could you? She'd done that before and the price she'd been given had stopped her from properly living.

When she returned to the Louvre, Madeleine found Véronique in her dressing room, packing up her doll and her automaton book, the portrait of her mother. The girl looked exhausted – her movements slow, her lips pale. Though she tried to hide it, she was still not recovered from the pain of the previous weeks. She would always walk with a limp.

'Here, let me do that, Miss Véronique. We'll wrap it in a cotton sheet.'

'Thank you, Madeleine.' Véronique sat heavily on her chair. 'Who'd have thought there'd be so much to do?'

Madeleine began removing the clothing from the wardrobe and folding it, while Véronique sat with her chin on her hands. After a minute, Véronique opened her dressing table drawer and took out the oval golden box. She inserted the tiny golden key, wound it three times and watched as the little girl – the rope dancer – jumped up and down on her perch. Madeleine moved to stand next to her mistress, her hand on the back of her chair. The two of them remained there together, watching, listening, until the mechanism gradually slowed, the music grew discordant, and the girl stood silent on her branch, ready to dance again.

Historical Note

The Clockwork Girl is very much a work of fiction. It was, however, partly inspired by real events, in particular the scandal of the Vanishing Children of Paris of 1750. When children began to go missing from the streets, various theories sprung up as to who or what might be taking them. In May, a lawyer named Barbier wrote in his diary: 'For a week now people have been saying that police constables in disguise are roaming around various quarters of Paris, abducting children, boys and girls from five or six years old to ten or more, and loading them into the carriages which they have ready waiting nearby.' Some said the children were being sent off to the colonies or to the wars. Some believed the police were using them to extract ransom money. Others thought something far darker was at work. Barbier noted the belief that the kidnappers were agents of 'a leprous prince whose cure required a bath in human blood, and there being no blood purer than that of children, these were seized so as to be bled from all their limbs'. And some said this prince was in fact their king. 'The wicked people ... are calling me a Herod,' whined Louis XV, no longer the *bien aimé*.

Following widespread riots and a public inquiry, it was determined that the real source of the vanishings was a royal edict of November 1749, which had directed that: 'all beggars and vagrants found in the streets of Paris … of whatever age or sex, shall be arrested and taken to prison, there to be detained for as long as shall be deemed necessary.' Lieutenant General Berryer, the man in charge of enforcing the edict, had wanted instant, tangible results. His harsh enforcement measures resulted in not just vagrants, but the sons and daughters of tradespeople running errands, and children playing in the street, being carted off in shuttered carriages and left in Paris's bleak houses of detention.

Louis XV did not, so far as I know, have a surgeon who carried out macabre experiments, but he did have an obsession with death (from which Pompadour tried to distract him) and a love of automata. He was known to have tasked master automaton-maker Jacques de Vaucanson with creating a moving, bleeding human for the purposes of scientific advancement. Vaucanson, however, never achieved that aim, so I gave the task instead to the fictional Doctor Reinhart.

Reinhart is partly modelled on the genius Vaucanson and partly on the philosopher René Descartes, who was said to have recreated his dead daughter, Francine, as a moving clockwork doll. According to the (probably apocryphal) story, the doll came to a watery end after the captain and crew of the boat carrying Descartes entered his cabin. There in a casket they found the mechanical doll and, believing her to be the work of witchcraft and evil, threw her into the sea.

Madame de Pompadour kept her place at court and remained an influential advisor to the King and a patron of the arts until her death from tuberculosis at the age of forty-two.

Glossary of Historical and Slang Terms

Academie des sciences –- The French Academy of Sciences

d'affranchir – to save a certain card at the cost of another

le bas peuple – the populace, the lower classes

beurre demi-sel – a girl or woman well on the way to becoming a prostitute

Le Bien-Aimé – The Well Beloved (name for Louis VX before his popularity plummeted)

belle mignonne – beribboned skull supposedly kept by Marie Leszczynska

bon sang – Good God

bonne affaire – bargain

cabinet noir – government intelligence gathering office

chambre ardente – special tribunal established for the trial of heretics

cocotte – prostitute (slang)

connard – asshole

culs-de-jatte – limbless beggars

dégueulasse – disgusting

femme entretenue – mistress / kept woman

femme de chambre – chambermaid

femmes de terrain – low prostitute

fichaise – a worthless thing, 'not worth a curse'

filles publiques – prostitutes

gens de qualité – upper classes

grisette – a young working-class woman

haute bourgeois – upper middle class

hôtel particulier – a grand townhouse

Lever - The Levée or Lever was the ceremony that took place every time the King woke up

macquerelles – brothel-keepers

maitresse en titre – the chief mistress of the king of France

mouches – police spies

ordre de cachet – *un letterre* or *order de cachet* was a letter signed by the king used primarily to authorise someone's imprisonment.

petits apartments – the king's private apartments at Versailles

pipe (faire une pipe) – to give a blowjob

poupée - puppet

pucelages - maidenhoods

quequette – prick

rousses – slang term for police

sérails – Paris brothel

tous azimuts – all over the place, in disorder

English words:

blunt – money

ceruse – white lead make-up

mort – woman

ottomised - anatomised

Acknowledgements

Some people are capable of shutting themselves away and writing a book entirely on their own. I am not one of those people. I'm very grateful to all those who've supported and inspired me over the years it took me to research and write *The Clockwork Girl*.

In particular, thank you to the following:

My brilliant agent, Juliet Mushens, and also Liza De Block and the Mushens Ent Team for their unstinting support and positivity.

My wonderful editor, Charlotte Mursell (and before her Olivia Barber), and everyone at Orion for believing in this book.

Micaela Alcaino for the stunning cover design.

All those who generously gave their time and expertise to assist me with my research, in particular Rupert Parsons, Matthew Read, Sabrina Bowen, Marisa Haetz, Tom Wedgwood, and the staff at the British Library. Any departures from fact – either intentional or accidental – are my responsibility alone.

My family for putting up with my nonsense year after year. In particular to my sister, Laura, and my husband, Jake, for reading and making inspired suggestions. Thank you also to my children for their amazing ideas, book cover designs, and constant distraction techniques.

All my writer friends, including North London Writers, South London Writers, Essie Fox, Mary Chamberlain, Colin Slong and The Lady Killers, for their advice, input and support.

Lastly, thank you also to the readers who have emailed or messaged me over the past few years, who have left reviews, blogged, tweeted, told their friends, or discussed my novels at book clubs. Do join up to my mailing list or find me on social media (which is where I often am when I should be writing).

http://annamazzola.com
https://twitter.com/Anna_Mazz
https://www.facebook.com/AnnaMazzolaWriter/
https://www.instagram.com/annamazzolawriter/

Credits

Anna Mazzola and Orion Fiction would like to thank everyone at Orion who worked on the publication of *The Clockwork Girl* in the UK.

Editorial
Olivia Barber
Charlotte Mursell
Sanah Ahmed

Copyeditor
Sally Partington

Proofreader
Clare Wallis

Audio
Paul Stark
Jake Alderson

Contracts
Anne Goddard
Humayra Ahmed
Ellie Bowker

Design
Tomás Almeida
Joanna Ridley
Nick May

Editorial Management
Charlie Panayiotou
Jane Hughes
Bartley Shaw
Tamara Morriss

Finance
Jasdip Nandra
Afeera Ahmed
Elizabeth Beaumont
Sue Baker

Marketing
Brittany Sankey